KU-323-704

INTERIBERICA, S.A. DE EDICIONES

PLANET EARTH

Earth's Atmosphere and Crust

Doubleday and Company Inc.,
Garden City, New York, 1977
A Windfall Book

SBN: 385 11344 7
Library of Congress Catalog Card No: 75 13118

Also published in parts as The Air Around Us and The Earth's Crust

PLANET EARTH

Part 1
The Air Around Us

by John Sparks

ISBN: 84-382-0018-4. Dep. Legal: S.S. 600-1975
Printed and bound in Spain by
T.O.N.S.A., and Roner S.A.,
Crta de Irun, Km.12,450, Madrid–34

Series Coordinator Geoffrey Rogers
Art Director Frank Fry
Design Consultant Guenther Radtke
Editorial Consultants Donald Berwick
David Lambert
Series Consultant Malcolm Ross-Macdonald
Art Editor Susan Cook
Editor Allyson Fawcett
Copy Editors Maureen Cartwright
Damian Grint
Research Julia Hutt
Enid Moore

Contents: Part 1

Editorial Advisers

Foreword by David Attenborough

One of man's most astonishing achievements is to have pieced together, from numberless tiny fragments of evidence, the history of the planet earth back more than 1000 million years or so before man himself arrived upon it. To guide his thinking in this extraordinary enterprise, he began with two fundamental propositions: first, that if one layer of unfolded rock, for example limestone or sandstone, lies on top of another, the lower layer must have been deposited before the upper one and hence it is older; and second, that no processes other than those going on in the world today should be used to explain the events of the past.

These two principles may seem almost absurdly obvious, but they were not always either recognized or accepted. Only a few centuries ago, acts of supernatural intervention were freely proposed in order to explain the shells and bones that people found embedded in the rocks, whereas the new reasoning seemed sometimes to lead to absurdities: if fossil shells were really the remains of creatures that once lived in the sea, and the limestone surrounding them the mud that accumulated on the sea floor, how could that limestone come to form the top of a mountain?

The realization that the earth was vastly older than anyone had hitherto supposed was one of the first steps in understanding such baffling questions. One of the last was taken only in the past few decades, when geophysicists identified the forces, still in action today, that have, over millennia, caused the continents to wander over the face of the earth and so crushed and buckled the rock basins of the sea that they have become mountain ranges.

Now, after little more than a century of research, scientists have unraveled many of the mysteries, and today it is possible for Dr. Bourne to sketch, in a mere 144 pages, a coherent biography of the planet earth. The story, even as it is understood by those researchers devoting their lives to its study, still has immense gaps. There are long episodes that can be—and are—interpreted in different ways. Without question, there are some that at the moment we misunderstand. It is a story that certainly will never be fully told, yet its broad outline is clear—and enthralling.

In the story of life there is no single hero. One by one, different forms of life have colonized different habitats, dominated them, and then, often, been superseded by others. As Dr. Bourne explains, life started in the sea and moved slowly onto the land, and eventually some creatures broke free of the ground and took to the air.

Dr. Sparks takes up the story of this last and fascinating enterprise. He describes in detail the surprisingly complex structure of the gas envelope that surrounds the world, and chronicles the ways in which life has managed to colonize the very lowest layers. This process has happened several times, and each animal group that has managed it has developed its own methods of reducing the effect of the pull of gravity on its body.

The insects were first. A group descended from reptiles were next—they grew feathers and became birds. Several different groups of mammals managed it subsequently, and, within the last few decades, man. First he rivaled the soaring of the birds. Then he pushed upward right through the atmosphere to become the first living creature we know of to leave completely the world that reared him, planet earth.

David Attenborough.

Breath of Life

It is easy to forget about the air around us. Because it is a mixture of gases, it cannot be observed, handled, and examined in the same way as lumps of rock or bodies of water. As it is largely out of sight, therefore, it also tends to be out of mind. Yet the thin atmospheric mantle of our planet exerts an influence out of all proportion to its visibility, or even its actual weight. Indeed, the layer of gas that accounts for less than one millionth of the earth's mass is crucial to the very survival of life.

One way of supporting this statement is to try to imagine what the earth would be like if, by some cosmic magic, the gases that form the basis of our atmosphere were dispersed into space instead of being hugged to the surface by gravity. We do not have to look farther than the moon to see some of the consequences. There, the force of gravity, which is only one sixth of ours, is insufficient to hold on to an atmosphere for any length of time. So the moon is airless, and much

of what we find strange about moonscapes stems from the absence of a blanket of gases.

Sunshine on the moon is nothing to sing about, because the surface receives the violent glare of the sun, unimpeded by an atmospheric filter. Thus unprotected, the moon receives not only the intense light of the sun, but also its full spectrum of shortwave electromagnetic radiation, such as ultraviolet and X rays. A steady stream of charged particles (the solar wind) together with highly energetic rays from outer space (cosmic rays) also assail the moon with deadly intensity. Life as we know it could not survive unprotected against such a high level of solar and cosmic radiation.

The moon's extremes of heat and cold are also lethal. Exposure to the merciless light of the sun causes the daytime temperature to soar to around 180°F—almost high enough to simmer a stew on earth. By contrast, the long moon night is much colder than the coldest earthly grave, for the

A giant "pinwheel" weather system revolving above the Pacific Ocean, as photographed by Apollo 9. Clouds—collections of water droplets and ice particles suspended in the air—are visible reminders of the invisible atmosphere that cloaks our earth.

daytime heat generated in the rocks is quickly radiated out into space after sunset, with no hindrance from a muffler of gases. So nighttime cold can reach an intolerable −220°F. Only our cosy atmospheric blanket shields us from such extreme temperatures.

Finally, the moon is exposed to a high level of meteoric bombardment. As the moon travels through space, pieces of interstellar debris are pulled toward it by gravity. With nothing to hinder them, they cannon into the surface at a speed of perhaps five miles a second. Traveling at this velocity, even a speck of dust contains a lot of energy, which is released on impact; large meteorites bury themselves in the ground and give up their energy of motion in a destructive explosion of molten rock. The whole moonscape shows evidence of this eternal bombardment, for the mountains have been eroded into rounded mounds by the abrasive impact of meteorites.

Luckily for us on earth, the air serves as a defense against this kind of blitz. As meteors plunge through space toward our planet, they collide with the rarefied outer layers of the atmosphere at speeds of more than eight miles a second (these high speeds are a result of the earth having a stronger gravity than the moon). Friction with the gas slows them down, and at the same time generates immense heat; as the temperature of the rock-missiles rises, they burn, and the smaller ones—called shooting stars—dissipate in a momentary streak of light long before they can reach the ground. (It is to guard against the danger of burning up in this fashion that the reentry capsules of space missions are equipped with ceramic heat shields.) Only a very occasional giant meteorite manages to get through the thick lower layers of air and crashes into the earth.

Living comfortably beneath our atmospheric blanket, we and all other living things on earth are insulated against the destructive effects of both our own star (the sun) and space. But the atmosphere makes its presence felt in many other ways, some of them very subtle. Although, apart from the clouds, we cannot really see the air, the sky looks blue because the air selectively scatters part of the spectrum of sunlight. Unless there is a wind blowing, we have no sense of being surrounded by air; yet it bears down on us with considerable force. Evidence for this was provided nearly 350 years ago in 1643, when an Italian, Evangelista Torricelli, demonstrated just how heavy the air is.

In the famous experiment, Torricelli filled a long tube—closed at one end—with mercury, which is a dense liquid metal at room temperature. Then he turned the tube upside down, making sure that none of the liquid was lost in the process, and submerged the open end in a bowl filled with mercury. The level in the tube slumped, but by only a small amount; a column of mercury about 30 inches high remained. It was supported, he concluded, by the weight of air pressing down on the mercury in the bowl. This conclusion was correct and Torricelli had thus invented the barometer, which, in its simplest form, measures atmospheric pressure by the rise or fall of the level of mercury in a tube.

We still measure atmospheric pressure in terms of inches of mercury. At sea level the air pressure is usually sufficient to support a column of mercury about 30 inches tall, which is equivalent to 14.1 pounds per square inch. This is the weight of the air above us; the total pressure of air upon our bodies is something of the order of 10 tons.

Of course, the pressure inside our bodies pushing out is more or less equal and opposed to atmospheric pressure, and so we do not feel the weight of air. Nevertheless, we are all animals who must manage to survive on an "ocean" floor. But whereas conditions on the actual seabed are fairly stable, the state of the air at the bottom of the atmosphere—where we live—varies from time to time, and this has a profound impact upon us.

The atmosphere is in a constant turmoil, which helps to brew the weather. We continually worry and talk about the weather; we follow eagerly the unfolding patterns of depressions, cyclones, and their associated fronts in daily weather charts. This is hardly surprising, because our daily routines are regulated by day-to-day changes in the weather. Not only do we search out warm sunshine whenever we can afford to, but we do everything we can to keep ourselves comfortable in bad weather. A great deal of industrial effort goes into helping us adjust to the weather and climate by means of umbrellas or sunshades, heavy coats or bikinis, air conditioners or heaters.

Less obviously, our moods and feelings are vul-

Only the life-support systems of Edwin Eugene Aldrin stood between this second man on the moon and its hostile, airless environment. To be robbed suddenly of its atmosphere, the earth would soon resemble the moon. The sky would be black. Lethal space particles would bombard us. We should bake by day and freeze by night. But long before that we would suffocate.

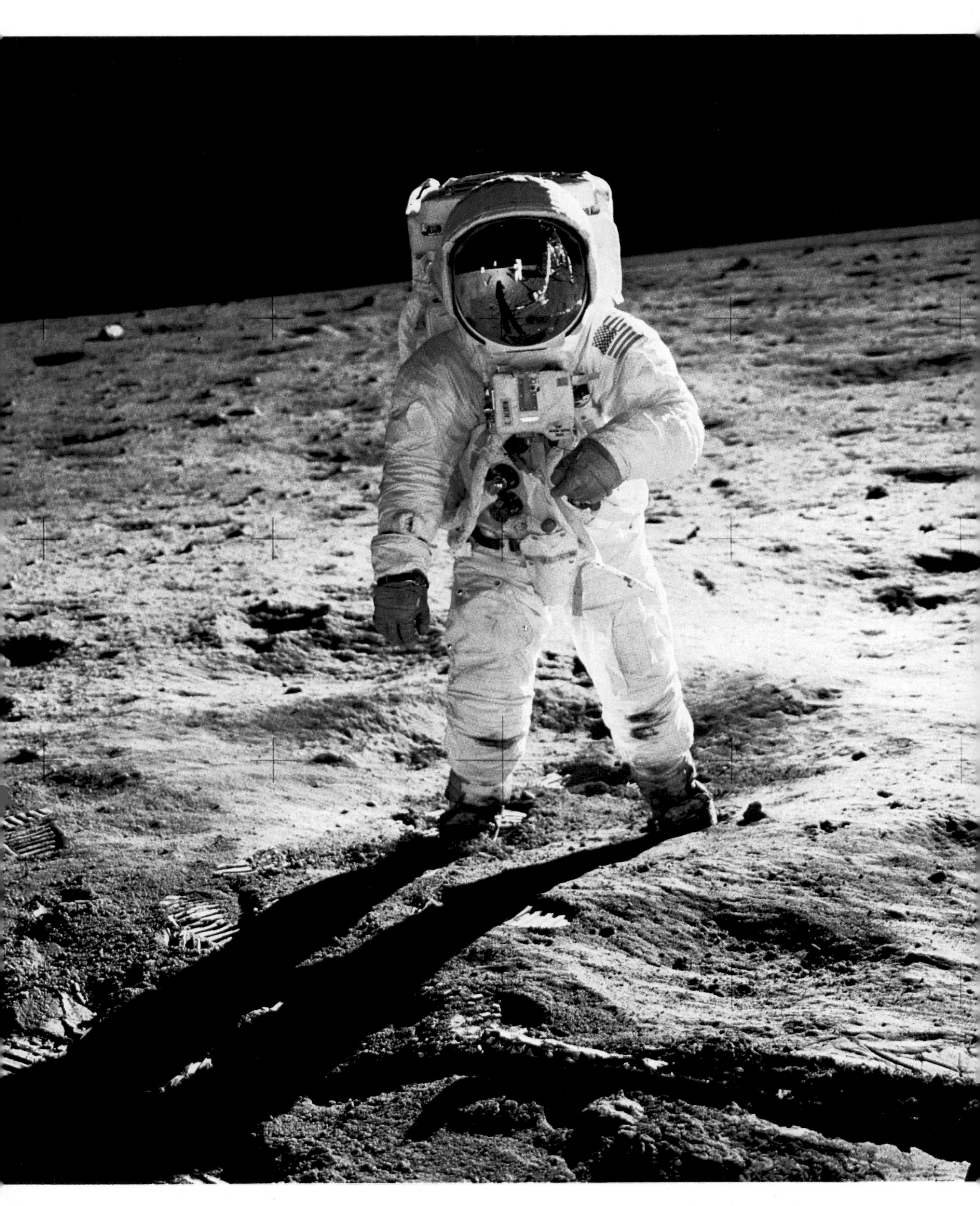

Left: seen through the mellowing filter of the earth's atmosphere, the setting sun appears as a benevolent ball of warmth. The air shuts out most of the sun's lethal radiation, yet traps solar heat, thus helping to stop the land freezing over at night.

nerable to climatic and atmospheric changes. Bracing dry air can be stimulating; humid conditions can lead to instability, irritability, or even extreme mental depression. Those responsible for law enforcement in America's big cities dread a hot, sultry summer, when they can expect civil violence to erupt. Not long ago, a criminologist analyzing the causes of assault and battery pointed out that the frequency with which these crimes are reported in the United States seems to vary with the temperature, rising to a peak in July. A drop in pressure is reputed to arouse pain in arthritic joints. And many people feel rundown in the wake of cold, wet weather.

Sometimes the turmoil of our atmosphere has a direct bearing upon survival itself. As we all know, weather disasters can strike with concentrated fury almost anywhere at any time. In the United States the annual cost of weather-induced damage varies from about 3000 million dollars to 10,000 million dollars; a few years ago, just one tropical storm did 1400 million dollars' worth of damage. And a single typhoon in Japan in 1959 killed 4600 people and injured eight times as many. But although it is such monumental tragedies that hit the headlines, the long-drawn-out effects of unseasonable weather are probably a greater menace. Sharp frosts destroy fruit in the making, dry skies wilt cereals, and high winds and rain flatten and rot crops awaiting the harvesters. It is little wonder that the nations of the world spend close to 2000 million dollars annually on efforts to predict weather conditions.

Weather is not merely an agency of doom and gloom, however. The air and all its manifestations add up to a beautifully balanced and unique life support system—unique, at least, in our solar system. Not that earth alone possesses an atmosphere. Far from it; most of the sun's other planets are shrouded in gas and vapors. But none of them has air like ours. They tend to be rich in hydrogen. For example, Jupiter has a thick atmosphere composed of free hydrogen, along with ammonia and methane, which are hydrogenated forms of nitrogen and carbon respectively.

Our air is unique because it contains plenty of oxygen—an element that is rarely found free in large quantities. Our oxygen-rich atmosphere would immediately focus the attention of a scientific observer from outer space, for he would conclude that something peculiar must be going on down here to produce so much of this highly active gas. He would be right. What is "going on down here" is life, and biological processes have considerably modified the air around us. A few thousand million years ago, our planet, like the others, was invested with a hydrogenated atmosphere, but living organisms have slowly conver-

An artist's reconstruction of Evangelista Torricelli's experiment. Air pressing down on the mercury in the open bowl supports some 30 inches of mercury in the sealed-top tube. Mercury that fell from the top of the tube has left a vacuum there.

ted it to a relatively oxygen-rich one, and plants are maintaining it even today.

In the following chapters we shall see how the atmosphere has affected the fundamental design and nature of animals and plants. It is easy to appreciate how the streamlined form of a fast-swimming mackerel is an evolutionary solution to reducing the energy needed for cutting through water. In the same way, air makes certain structural demands on any organism that exploits flying as a means of locomotion. Wings, whether of bee, bird, or sycamore seed, must obviously meet those requirements. But, as this book will show, the need to come to terms with air expresses itself in many less obvious ways.

Local changes in atmospheric pressure, temperature, and humidity give rise to the rich variety of weather that can sometimes bless, and sometimes assail, the world. Above: caricature of 1808 by British artist James Gillray stresses the discomfort of windy weather. Below: snowbound automobiles at Grand Prairie, Alberta, western Canada, remind us that even our modern transportation systems are no match for the weather. Right: monsoon rains inconvenience Bombay's umbrella-bearing citizens but will revive the parched fields that feed them.

The Ocean of Air

How deep is the ocean and how high is the sky? The first question is much easier to answer than the second. The sea has a distinct surface, and we can measure its depth at any point over the seabed. But the gaseous ocean above us has no clear-cut boundary with space. With increasing altitude, the air becomes rapidly thinner, and its pressure drops. The nature of gas is responsible for this sharp decline in density with height. Unlike water, gas is very compressible: in other words, a given mass can be made to occupy less volume if pressure is applied, or expanded if the pressure is reduced. It follows that the atmosphere is densest where it bears the greatest weight, and this is at ground level, where the pressure of the overlying atmosphere reaches a maximum. The 17th-century French mathematician Blaise Pascal proved this when he sent his brother for a walk 3200 feet up a mountain with a barometer; the column of quicksilver slumped three inches because of the decreased weight of the overlying air. Had he been able to ascend higher, he would have found a continuous fall in his barometer as the air became increasingly more rarefied.

In theory, our planet's atmosphere could stretch as far as the point at which its density and pressure fall to the near vacuum of space. A century ago, the belief was that this happens about 100 miles up, and the atmosphere was accordingly visualized as a thin film of gas adhering to the hard globe of the earth. We know today that this notion falls far short of the truth. Yet in one sense it is correct. Although the mass of our atmosphere is reckoned to be a stupendous 5600 million million tons, 80 per cent of it is compressed within a layer only 5 to 10 miles thick, and all but 1 per cent is below 19 miles. Thus the greatest slice of the atmosphere is held close to the surface of the earth. But space exploration over the last couple of decades has revealed a detailed picture, showing not only how complex it is, but also the surprising fact that it can be

A vast cloud of volcanic particles rises high in an evening sky gleaming with sunlight scattered by airborne dust. Man marveled at such dramatic scenes long before he understood anything about the ocean of air that made them possible.

detected for thousands of miles into space.

The gaseous mantle of our planet is divided into a series of more or less defined layers, each with its own characteristics. We call these layers the *troposphere,* the *stratosphere,* the *mesosphere,* the *ionosphere,* and the *exosphere.*

The troposphere, which is the lowest and densest layer, varies from 6 to 10 miles thick. Broadly speaking, it contains the bulk of the air and most of the atmospheric water, and it is turbulent, with powerful winds and convection currents spawning the weather. From ground level to its ceiling—or *tropopause*—there is a gradual drop in temperature. In fact the temperature decreases by approximately 3.8°F for every 1000-foot increase in altitude. This may seem rather odd, because the upper levels of air are nearer the sun. What actually happens is that, in the absence of highly reflective and opaque clouds, the air of the troposphere allows nearly all the visible sunlight through to the ground. There, some of it is absorbed and the energy converted into heat, which raises the ground temperature. Some of this heat is transferred directly to the air by conduction and convection currents. But the warm surface also radiates some of this energy back into space by emitting infrared rays, to which the troposphere is not transparent. The outward-going radiation is therefore absorbed mostly in the lower regions of the troposphere, with a consequent warming up of the air. This is a major reason why the air is warmer near the ground than at higher altitudes.

It was once thought that the decrease in temperature could give a clue as to how far the atmosphere extended. It had been assumed that the relationship between temperature and altitude was maintained in a steady fashion until the temperature of the ever-thinning air reached absolute zero; in theory, this should happen at a height of 20 miles, which would mark the limit of the atmosphere with the frigid interstellar void. The assumption was disproved at the end of the 19th century, when a French scientist, Léon Teisserenc de Bort, sent balloons with recording

Left: lightning strikes a manned balloon caught in a rainstorm—an artist's impression of a disaster of the early 1890s. Atmospheric turbulence and shortage of breathable air at high altitudes discouraged manned exploration of the ocean of air. Instead, scientists began designing instrument-carrying robot balloons, which sounded out the upper levels of the atmosphere.

Right: from an earth-orbiting satellite many miles up we look down through the invisible layers of air to Egypt's Nile Valley (green strip) with the Red Sea and Arabia beyond. Man-made satellites give us new insight into the earth's atmosphere. For example, this photograph reveals the extent of a long streak of high-altitude cloud aligned east-west and marking the lower edge of a jet stream. Such currents of air flow up to 400 miles an hour and occur when air masses of different temperatures meet.

equipment to hitherto unexplored heights. The instruments showed that at an altitude of six or seven miles over Europe the temperature stopped decreasing and started to increase. This discontinuity, which marks the ceiling of the troposphere is a universal characteristic of the atmosphere, although it varies in height from about 6 miles over the poles to about 10 miles over the equator.

Beyond the tropopause is the relatively stable stratosphere, which rises for about 20 miles above the tropopause. Convection currents carrying heat from the earth's sun-warmed surface rarely rise into it, and yet the temperature rises. At a height of 30 miles above sea level, it may be around 32°F—some 90°F warmer than the temperature of the tropopause. The heat is generated partly by a process crucial to the survival of life down below: the formation of ozone from oxygen. Ozone is a highly active and lethal form of oxygen, and it is fortunate for living forms that the traces of it in the atmosphere are chiefly confined to the upper portion of the stratosphere, where its presence is actually beneficial to us. At these heights it is formed by a collision of the sun's high-energy ultraviolet radiation with oxygen molecules, splitting them into their constituent atoms. These atoms in turn collide with, and hitch up to, other oxygen molecules to form the denser gas, ozone.

The amount of ozone present in the atmosphere at any one time is relatively small. At sea-level pressure, there would be just enough to form a layer one hundredth of an inch thick. But the ozone in the stratosphere, with a peak concentration 22 miles up, acts as an effective barrier to the worst excesses of the sun's intense ultraviolet emissions. Thus this dangerous gas does us good by shielding the surface of our planet from another great danger, for the ultraviolet that the dense ozone screens off has the unpleasant ability to plow through tissues and disrupt DNA—the genetic material in our cells—with disastrous consequences. Some of the ultraviolet does penetrate to ground level, of course, as pale-skinned people who suffer from sunburn will

Molecules of air, dust, and water vapor in the atmosphere scatter some of the radiation beamed down upon the earth from the sun. Blue light in particular is scattered, which is why the black sky appears blue when seen from earth.

Much of the incoming solar radiation (shown here by arrowed lines) almost reaches the earth's surface before it hits clouds and is reflected into space as "earthshine." One-quarter of all solar radiation is lost to earth by clouds.

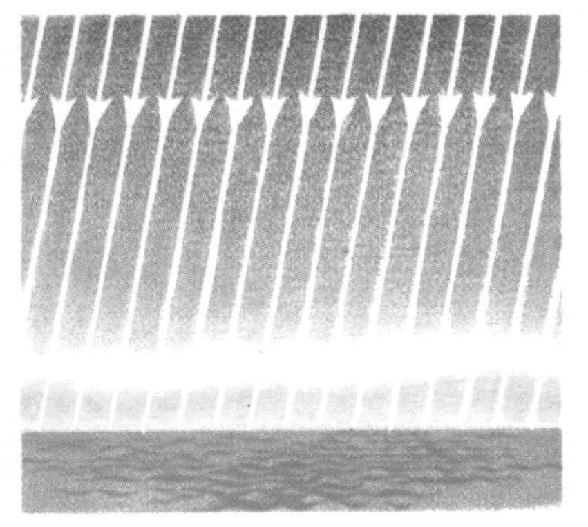

Less than half of all incoming radiation reaches the earth. But this fraction, transformed into heat, is enough to warm sea and land and the lower atmosphere. By making water evaporate and air move about, this heat helps to create weather.

painfully recall. But the ozone serves as a virtually opaque barrier to the most damaging radiation. Above the stratopause is a layer about 20 miles thick called the mesosphere, although some sources categorize it as simply the upper portion of the stratosphere.

The next layer to affect our lives in one way or another lies between about 50 and 300 miles up and is called the ionosphere. At this distance the air pressure is one millionth of that recorded at ground level, but it is higher than the near-vacuum of space. What little gas there is in the ionosphere is excited by the savage intensity of X rays and ultraviolet. The atoms and molecules absorb some of the radiation, which heats them up; many are shaken apart in the process, and produce electrically charged fragments, or *ions*. Two hundred miles above the earth's surface, the density of negatively charged electrons that have been stripped from gas molecules reaches 15 million per cubic inch! And they are of some use to us because they make the ionosphere behave like a mirror to radio waves, and thus permit radio signals to be reflected around the world.

Absorption of shortwave radiation also causes a steady increase in temperature. At a height of 185 miles it is about 2400°F, and farther out it reaches about 6840°F, which is very hot indeed. But of course the gas is so thin at such heights that the amount of heat it contains is correspondingly minute. This may sound like a contradiction, but high temperature does not necessarily produce great heat. The heat at a particular temperature is related to the amount of substance present. Consider a spark at several hundred degrees Fahrenheit—which could fall on you and not feel hot at all—and compare this with the scalding heat of a bath of water, the temperature of which is below boiling point.

Sunbathers—one pale, one tanned—on Cannes beach in southern France. The ultraviolet rays from the sun that penetrate the atmosphere burn skins that are unprotected by pigmentation.

Although the ionosphere reaches well beyond the orbits of many man-made satellites, it does not mark the limit of the earth's atmospheric mantle. The outermost zone, or exosphere, begins at a height of about 300 miles. Beyond this point, the gas molecules are so thinly spaced that they rarely collide. At ground level, gas molecules travel at 1000 miles an hour and collide and ricochet 5000 million times a second; but in the exosphere many follow ballistic courses, soaring out into space then falling back under the pull of gravity. Some are literally in orbit, whereas others, boosted by collisions, reach escape velocity, break away from the earth's gravity, and zoom out into space.

It is probable that the final frontier between our planet and space is not hundreds but millions of miles away. In recent years we have learned that the earth is almost totally surrounded by a series of concentric radiation belts, which are known as *Van Allen belts* (after the American physicist who was chiefly responsible for their discovery). In these huge layers, thin ionized gas gyrates backward and forward along the magnetic lines of force between the northern and southern latitudes. The earth's radiation belts, together with its magnetic field—what scientists call the *magnetosphere*—reach out about 50,000 miles toward the sun and extend about 3½ million miles on the dark side of the earth. But the outer regions of the magnetosphere, beyond the Van Allen belts, interact with the sun's own atmosphere, for although the sun is nearly 93 million miles away, the hot ionized gas that streams from it extends well beyond the earth's orbit. Some of this solar wind is attracted by our planet's magnetism into the Van Allen belts. Only at the immense distance of roughly 3½ million miles from the earth's surface do the charged particles in the solar wind swamp the weak magnetism of the far-off earth. And this, we might say, is where space really begins.

But to return to the ocean of air surrounding our planet, let us now consider the composition of that ocean. The atmosphere is a mixture of gases and vapor, and its composition is remarkably constant up to great heights. Ignoring for a moment dust particles, spores, seeds, flying animals, and a whole spectrum of industrial effluents that are liberated into the air, the major ingredients are nitrogen and oxygen. In dry air, they occur approximately in the proportion of four parts of nitrogen to one of oxygen.

The discovery of the constituents of the air around us is a relatively recent one. Something over three centuries ago, a new philosophy was gaining ground in Europe, based upon the method of making testable deductions from experimental observations. We now take this scientific method for granted, but during the 15th and 16th centuries, thinkers with open minds were only just beginning to question medieval notions, which were usually firmly rooted in tradition and dogma. Alchemists, with their objective of turning all that did not glitter into gold, reigned supreme, and it was generally believed that everything consists of four "elements"—earth, air, water, and fire. It took a very long time to establish that there are over 100 natural elements.

The nature of gases or "air" was especially difficult to investigate and remained so for a century or more. At the time, gases could not be seen, packaged, weighed, liquefied, or solidified. Knowledge about them remained essentially speculative until sophisticated scientific tools became available in the 17th century. A Flemish chemist, Jan Baptista van Helmont, coined the term "gas" in the early 17th century: he related his new word to the Greek word "chaos," meaning "something unformed and without shape," and he described gas as something that could not be kept in a vessel. He was the first to realize that there were different gases, that the "substance" of air—oxygen—was one of them, and that carbon dioxide was another (he called it, appropriately, *gas sylvestre*—"gas of the woods"). And he performed an interesting experiment, although he was unable to interpret accurately what he saw. He floated a burning candle in a water trough, and then inverted a jar over it. Soon the flame went out, and at the same time the level of water in the jar rose. Because the burning candle had obviously reduced the volume of air in the jar, he thought that the flame had annihilated something. So it had, in a way. In fact, it had used up all the oxygen, and the carbon dioxide that had been formed dissolved in the water.

Another chemist, the Englishman John Mayow (1641–79), carried van Helmont's experiments farther: he introduced a mouse into the bell jar after the candle had become extinguished, and it promptly suffocated. Furthermore, he discovered that a mouse placed in a bell jar over water along with a lighted lamp lived for only half as long as the mouse by itself. From his observations he more or less correctly deduced that air is a mixture of two gases, one of which supports combustion and respiration (and which he called "nitrous aerial spirit"), the other of which is incapable of supporting either. It took more than 100 years for Mayow's deduction to become an established fact.

At last in 1772, a German-born Swedish apothecary, Karl Scheele, managed to isolate Mayow's "nitrous aerial spirit." He noted, too, that if he mixed this gas, which he called "fire air" (oxygen), with "foul air" (nitrogen) in the proportion of one to four, the resulting gas was indistinguishable from the air we breathe. The search for the nature of the atmosphere was also carried forward during the latter years of the 18th century by such notable scientists as Britain's Joseph Priestley and the great French chemist Antoine Lavoisier. And so, by the end of the

Five main "slabs" make up the layer cake of the earth's atmosphere. The turbulent troposphere breeds clouds and weather. In the stratosphere, ozone shields earth from space radiation. The temperature here is high but drops in the mesosphere. It is in the ionosphere that meteors burn up. Here, and in the exosphere, solar particles hit atoms and molecules in air, creating auroras—the lights that are often seen high in polar skies.

600 miles

Exosphere
300 miles

Ionosphere
50 miles

Mesosphere
30 miles

Stratosphere
6-10 miles

Troposphere

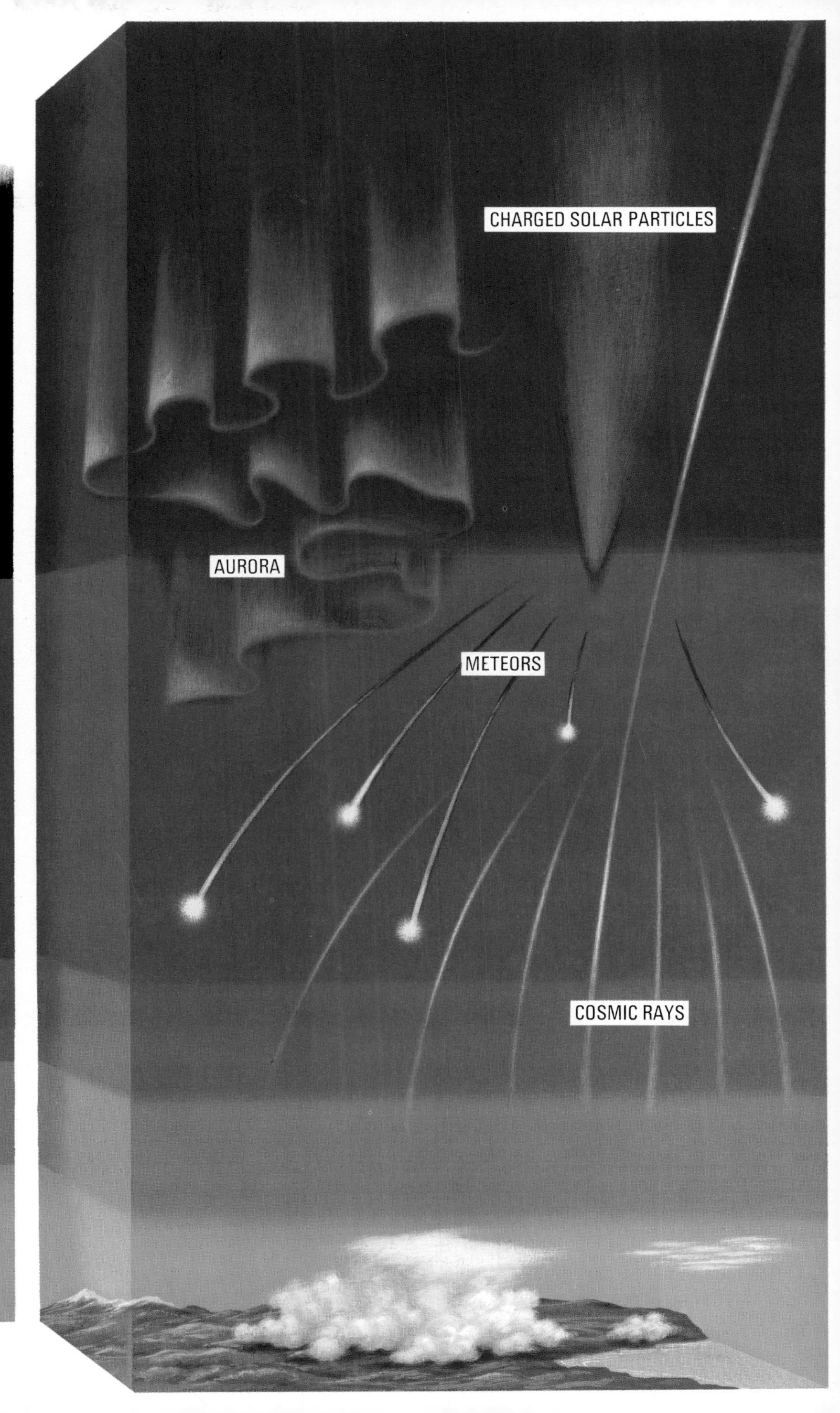

The ghostly ribbon of the northern lights (or aurora borealis) flickers more than 70 miles above Kiruna, northern Sweden. These glowing lights are produced by solar particles bombarding atoms and molecules present in the earth's upper atmosphere.

century, the general nature of air was known. We have learned since of a third element in the atmosphere: a fixed amount of an inert gas, argon, formed by the breakdown of radioactive elements in the earth's crust. But argon accounts for only 1.3 per cent of the mass of dry air.

There are, of course, other constituents of the atmosphere, but these occur in variable amounts from place to place and from time to time. Among them are carbon dioxide and water vapor. Although present in only small quantities, both play an enormously important part in moderating our climate and allowing life processes to continue. On average, carbon dioxide accounts for a mere three parts in 10,000 of air (0.03 per cent). In crowded towns and inadequately ventilated rooms, the carbon dioxide content may reach a much higher level. But continued breathing of concentrations as high as 20 parts per 10,000 (0.2 per cent) is injurious to health. Water vapor content is much more variable. It is, in fact, the most variable constituent of the atmosphere.

At earth temperatures, most water is either liquid or fossilized in mountain glaciers and polar ice caps. It evaporates from the surface of bodies of water and is transpired into the air from plants. The vapor in the atmosphere seldom rises higher than eight or nine miles and the amount present may vary from almost zero over some mountains and deserts to three or four per cent over humid tropical forests. This may not sound very much, but a small volume of air can carry a great deal of water. A 25-foot-square room with a 15-foot-high ceiling may contain up to 10 pounds of water vapor on a not particularly warm summer day. It has been calculated that if all the water in the air condensed out, it would form a layer one inch deep over the earth's surface at sea level.

The capacity of air to carry water increases as it warms up. When it contains the maximum capacity, it is said to be *saturated*, with a relative humidity of 100 per cent. Perfectly dry air would have a relative humidity of 0 per cent. The very driest climates on earth may have relative humidities as low as 5 to 10 per cent, whereas when rain is falling or it is foggy, the air is saturated. As saturated air cools, its carrying capacity drops, and the excess moisture condenses to form clouds, which in turn may precipitate as hail, snow, sleet, or just rain.

Although dull skies and damp weather can be miserable, it may be some consolation to reflect that the moisture in the atmosphere, together with the carbon dioxide, acts as a muffler to help keep the planet warm. As we have seen, the troposphere allows the sun's radiant energy through to warm up the ground. If air consisted of nitrogen and oxygen only, this warmth would be reradiated into space in the form of long-wave infrared; and with nothing to keep the heat down nighttime would be icy cold. It is the small quantities of water and carbon dioxide in the air that catch the long heat waves and return them to earth. Thus, carbon dioxide, water droplets in clouds, and vapor in the atmosphere serve the same purpose as the glass roof of a greenhouse.

In addition to nitrogen, oxygen, argon, carbon dioxide, and water vapor, the air also contains minute amounts of such other gases as oxides of nitrogen produced by lightning and other atmospheric phenomena, hydrogen sulfide, methane, and ammonia. As we all know, the atmosphere is profoundly influenced by the activities of living organisms—the *biosphere*. For instance, green plants release vast quantities of oxygen into the air; and animals in turn take in the oxygen and give out carbon dioxide. But how did the earth's atmosphere as we know it evolve in the first place? For the answer to that question we must turn the clock back several thousands of millions of years.

Sometime between 4000 and 5000 million years ago, the solar system was born and the earth formed out of what remained of the cloud of gas and debris that originally formed the sun. At first the planet may have been a sphere of silicate rocks and iron, but heat released from the decay of radioactive materials ultimately melted the heavy metals, and they sank to the interior, leaving the lighter silicates floating on the surface, where they formed a crust. Although devoid of life, the young world must have been a dramatic place, with the surface torn by

A dark sheet of cloud obscures the sky over these sunlit trees and fields. Water, in the form of clouds and water vapor, forms an important ingredient in the atmosphere.

powerful convection currents from within. Extensive volcanoes must have injected large volumes of nitrogen and steam into the atmosphere; and as the planet cooled, the steam condensed out to form the oceans.

For millions of years the sterile planet had a tropical, humid climate, with a highly poisonous atmosphere composed almost entirely of hydrogen, ammonia, and methane. Although oxygen was by far the most common element in the crust, only trivial amounts of the free element were present in the primeval atmosphere; we know this from the fact that rocks laid down at the time contain no oxidized metallic minerals—a sure sign that they were never exposed to oxygen. Nevertheless, the nitrogen- and carbon-rich atmosphere of 3500 million years ago did allow the first step to be taken in the origin of life. Organic molecules such as amino acids—the building blocks of proteins—can be produced in

Lightning bolts forge parallel paths through the atmosphere in a desert storm over Utah's Monument Valley. Lightning helps to produce oxides of nitrogen that are present in the air.

a mixture of methane, hydrogen, ammonia, and water by energizing the mixture in some way. There was plenty of opportunity for such a reaction to take place in those days, with intense ultraviolet radiation from the sun reaching the earth's surface, fierce volcanoes spewing out scalding gases, and electrical discharges no less common than they are today.

Somehow, then, self-reproducing organisms evolved from the amino acid precursors, and they doubtless found a plentiful supply of organic food in the seas. The bacterialike organisms may have lived on this organic soup by fermenting their food in the same way as yeasts do today. In other words, without the help of oxygen they broke down complex organic compounds and made use of the energy thus released. This was, of course, an inefficient process; think how much energy is still left in alcohol, which is a waste product of the fermentation by yeasts of sugar! Another drawback was that these first organisms were living on a capital organic resource that would sooner or later run out. Nevertheless, because carbon dioxide is a waste product of fermentation, the activities of fermenting organisms may have started to modify the environment by adding carbon dioxide to it, and this may have paved the way toward the next stage in the evolution of life.

By about 3000 million years ago, an increasing accumulation of carbon dioxide in the seas and atmosphere provided an opportunity for the evolution of organisms that could manufacture their own organic molecules from it, thus no longer needing to depend on the dwindling resources of the sea. An organism that can build its own food out of inorganic materials is a producer (or *autotroph*), as opposed to a consumer (or *heterotroph*). Producers need a source of energy to be able to manufacture complicated organic molecules from simple molecules such as carbon dioxide. The first producers—like some bacteria today—may have powered their molecular building operations with the energy released from simple chemical reactions.

Green plants harness the power of sunlight. Essentially, they absorb some of the solar energy in the blue and red region of the spectrum by means of the green pigment chlorophyll, and they use this energy to dissociate water into its component parts of hydrogen and oxygen and to synthesize carbohydrate fuels, such as glucose, by adding the hydrogen to carbon dioxide. This process, known as *photosynthesis*, liberates the oxygen contained in the water. Thus, the oxygen that is such a vital ingredient of the air was originally dumped—and is still being dumped—by plants as waste produce.

The very first photosynthetic producers were

accustomed to living in a world devoid of free oxygen, and it must have taken them time to get used to an environment with oxygen in it. Many bacteria that live today by fermenting organic substances even find oxygen toxic at a relative proportion of as low as 1 per cent. And we also succumb to oxygen poisoning at high pressures. The problem with oxygen is that it is a very reactive element, and will combine only too easily with a whole range of substances, including the carbon-based compounds that form the basis of life on this planet. Just consider how eagerly wood and oxygen combine in a fire; and finely powdered carbon will catch fire spontaneously in pure oxygen. To a living organism, free oxygen is hot to handle unless the organism possesses the right chemical machinery to take it quickly out of circulation before it oxidizes and damages the delicate carbon-based framework of living cells.

So the first autotrophs that managed to wrench hydrogen from water molecules must have found the waste oxygen dangerous stuff, and powerful selection pressures must have operated on those organisms to render the gas harmless. When eventually the early cells had evolved protective mechanisms against waste oxygen, as typical photosynthesizing plants have obviously done, free oxygen could be tolerated, and the "waste" produced began to accumulate in the seas and to diffuse into the atmosphere. The air apparently began to become significantly oxygenated about 1800 million years ago.

It is difficult to say what forms life took in that distant past. The liberation of oxygen into the air may have come about through the photosynthetic activity of primitive blue-green algae. The fossil record has yielded the petrified remains of upside-down cones made by these marine plants and sediments trapped by their filamentous structure. Small versions of such cones exist on the Florida and Australian coasts, but

over 1000 million years ago similarly shaped structures of sand and algae must have loomed fifty feet above the seabed. Although they may have been responsible for raising the oxygen level in the atmosphere to only about 1 per cent of today's level, this would have been sufficient for the formation of an ozone screen, which could have sheltered both the land and the sunlit upper waters of the seas from destructive radiation. As a consequence, there may have been a further increase in *phytoplankton* (floating water plants) and enhanced oxygen production. Even today 90 per cent of the photosynthetic release of oxygen takes place in the sea.

The presence of ever-increasing amounts of free oxygen set the stage for a further advance: the evolution of animals capable of using this oxygen to "burn" the high-energy fuels synthesized by plants, changing them back into water and carbon dioxide. This process, known as *respiration,* can be regarded as the "reverse" of photosynthesis; in respiring, an animal removes hydrogen from organic molecules and, by means of a complicated series of steps facilitated by enzymes, combines it with oxygen, and forms water. The energy liberated during respiration is geared to the synthesis of high-energy compounds, which in turn can be made to drive many biochemical processes, including the manufacture of body-building substances and the production of mechanical work.

By 1000 million years ago there was enough oxygen around to sustain large numbers of tiny oxygen-breathing animals, which fed either on plants or on other creatures like themselves. But even when, according to the fossil record, multicellular animals entered the picture—about 600 million years ago, at the beginning of the Cambrian period—the atmosphere may still have contained only between 3 and 10 per cent of the present-day level of oxygen, which would be far too little to keep modern animals alive.

How the atmosphere has evolved since then is not known for sure. Clearly the amount of oxygen dumped into the air by plants has, on balance, exceeded the demand for respiration and ordinary chemical oxidation. This planet has seen several phases of great volcanic activity, which may have temporarily removed large quantities of oxygen, as freshly emitted carbon monoxide was oxidized to carbon dioxide.

Because free oxygen is a waste product of the photosynthetic manufacture of organic carbon compounds, the build-up of oxygen in the atmosphere must mean that carbon is being taken out of circulation. Much of our earth's carbon was "banked" during a warm damp period, starting about 345 million years ago, which has been appropriately named the Carboniferous period. At that time much of the land was covered in lush, swampy forests of tree ferns. There were as yet no birds singing in the foliage, no reptiles and mammals scurrying through the undergrowth. Insects had made their appearance, however, as had some amphibians. For 100 million years these great forests must have absorbed prodigious quantities of carbon dioxide from the atmosphere and converted much of it into woody tissues. But during the Carboniferous period a small pro-

Invisible gases, hot "steaming" water vapor, and molten rock spurt from a Hawaii volcano. In the remote past such emissions may have helped to energize the rich chemical mixture on the earth's surface so as to create the first organic molecules.

Millions of years ago plants extracted huge quantities of carbon from the atmosphere. Trapped underground, this carbon became a raw material for coal. Above: Noamara Indians in a Chaco rainforest—part of the immense tract of tropical forest that sprawls over much of South America, recalling the lush Carboniferous forests that gave rise to coal. Left: Iranian oil flares light up the night. They burn the gas emitted by underground reservoirs of petroleum, a fossil fuel rich in carbon, formed from millions of tiny organisms long extinct.

Above right: blue haze hangs over a eucalyptus forest on the eastern scarp slope of the Great Dividing Range in southeastern Australia. The haze is caused by aromatic oils exuded by the trees to cut down water loss. Vegetation annually adds hundreds of millions of tons of organic vapors to the atmosphere.

portion of plant remains must have sunk down into bogs and swamps, where little oxygen penetrated; these plant remains ultimately formed vast seams of coal, lignite, and asphalt, as well as oil shales and underground "seas" of petroleum. Another period of coal formation took place about 70 million years ago. It is believed that since then the oxygen level in the atmosphere has been quite steady.

There are other great reservoirs of both carbon dioxide and oxygen. Limestone and chalk, for instance, are both calcium carbonate, and enormously thick seams of this substance cross the surface of the earth. The English Channel has exposed part of a 900-foot-thick layer forming the white cliffs of Dover, built from skeletons of planktonic algae that settled to the bottom of a warm shallow sea more than 70 million years ago. The carbon and oxygen contained within such mineral deposits have been taken out of circulation. But when they were being formed they must have influenced the composition of the air.

You will already have realized that the constituents of the atmosphere are in circulation, passing regularly into living tissues and into rocks and water before being returned to the air. The round trip that any given element takes has its own special timetable. Water itself is constantly recycled; it has been estimated that all the earth's water is split into its constituent elements by plant cells during photosynthesis and reconstituted by animal respiration on an average of once every 2 million years—with part of the cycle spent in the atmosphere as vapor. Oxygen makes the journey from air to air 1000 times more frequently—once every 2000 years.

The carbon cycle is worth looking at in some detail. Carbon completes its round trip through the atmosphere, oceans, and biosphere comparatively fast—every 300 years. Expressed another way, there is enough carbon dioxide in the atmosphere and sea to allow plants to carry on photosynthesis at their present rate for 300 years. About 200,000 million tons of carbon dioxide are annually transformed into plant material through photosynthesis, mainly by the sea's phytoplankton. Some 10 per cent of this amount is taken from the atmosphere—mostly by forests, which are by far the chief consumers of atmospheric carbon dioxide, accounting for about two thirds of it. Nearly all of this carbon dioxide is locked away temporarily as wood in trees and this represents the most important store (apart from the fossil fuels) of biologically fixed carbon.

The carbon cycle is completed by the decaying process. When plants and animals die, their carbon-rich remains are eventually returned to the atmosphere as carbon dioxide, either directly or through the action of bacteria, fungi, and insects. Some of the more volatile organic substances are even liberated into the atmosphere, particularly when the plant foliage is dying. It has been reckoned that 440 million tons of these organic vapors are returned to the air every year; their concentration is sometimes so great that they form a blue haze above some forests.

Carbon dioxide is soluble in water, and there is a constant exchange of carbon dioxide between the atmosphere and the sea. Tests using radioactive carbon atoms indicate that a carbon dioxide molecule stays in the atmosphere for only 5 to 10 years before dissolving in the sea. This means that about 100,000 million tons of atmospheric carbon dioxide disappear into the sea each year and are replaced by a nearly equal amount of oceanic carbon dioxide.

Even atmospheric nitrogen flows through the biosphere. Although, in comparison with the other gases, nitrogen is inactive, it forms a vital part of our living molecular architecture. Plants need nitrogen, but before they can utilize the gas, it must be combined with either oxygen or hydrogen—a process that requires the expenditure of energy. A small but significant amount is energized by radiation or lightning, so that it can react with oxygen and change into one of the oxides of nitrogen, or can form ammonia by reacting with the hydrogen in water vapor. The annual production of nitrogen in the atmosphere

For thousands of years natural and man-made fires have poured smoke and carbon dioxide into the air. Left: pall of smoke from a bush fire near the (New York–New Jersey) Lincoln Tunnel. Above: cones of the jack pine, a fire-adapted species. Its seeds germinate only if the cones have been roasted by the heat of a forest fire.

made available to plants and washed into the soil by rain has been calculated at 8.8 million tons. By far the most important source of nitrogen to the biosphere is that which is captured and converted (*fixed*) by microorganisms living either free in the soil or in conjunction with plants. Each year about 48.4 million tons of nitrogen are biologically fixed in this way. And another 30 million tons are artificially fixed in the manufacture of fertilizers. To balance the nitrogen that is taken from the atmosphere, denitrifying bacteria are able to release nitrogen back into the air during the decaying process.

The atmosphere, then, is in a continuous state of flux. And over the last 200 or so years, its composition has been considerably—and rapidly—modified by the activities of man. Industrialization requires vast supplies of energy, and this has been generated by furnaces of one kind or another. Of course burning is new neither to man nor to nature. Fire started by lightning or other natural phenomena has been beneficial to wildlife communities from time immemorial, because it has been the natural way to clear large areas of choked forest, thus allowing fresh regeneration to take place. Some trees have become so adapted to fire that their cones will not open unless scorched; those of the jack pine, for example, release their seeds at 140°F, and seeds germinate best if the cones have been roasted. Slash-and-burn agriculture has been practiced for thousands of years. There is evidence that the African savannas have been fashioned by the periodic ravages of both natural and man-induced fire.

Industrial burning, however, differs in its scale and in the fact that the fires are fueled by fossil carbon. Nearly 300,000 million tons of coal are mined each year; the annual world production of crude oil stands at a staggering 20,000 million barrels, and the rate of production rises about 7 per cent a year, so that the amount of oil gushing from underground reservoirs doubles every 10 years. We are therefore annually returning between 5000 and 6000 million tons of carbon to the atmosphere, chiefly in the form of carbon dioxide. And this is carbon that was fixed and "banked" beneath the soil millions of years ago.

Clearly, an increase in the proportion of carbon dioxide in the air must be expected. And even though the oceans tend to buffer the effect (because there is always likely to be 60 times as much soluble carbon dioxide in the sea as in the atmosphere), a rise in carbon dioxide levels has indeed been detected. Since 1860 there has been a 10 per cent increase—from 2.9 parts per 10,000 to about 3.2 parts per 10,000—and the rate of increase is almost certain to rise. Scientific readings taken in the Antarctic, away from local peaks in industrial effluent, indicate that the level is rising at 0.007 parts per 10,000 each year. If these trends continue there may be as much as 4 parts per 10,000 by the end of the century, increasing to 5 parts per 10,000 by the year 2020!

Considering the amount of carbon dioxide at present being spewed into the air by our factories, homes, power stations, and cars, these increases may seem surprisingly low. One reason for this is that a lot of the gas has undoubtedly gone into the sea. Furthermore, much of it may be taken up in extra photosynthetic activity by land plants, because the rate at which they can fix carbon has always been limited by the small concentration of carbon dioxide in the air. The more there is available, the faster photosynthesis can proceed under constant conditions of light and temperature. After a certain point, though, the relationship would break down, because carbon dioxide poisoning would set in.

The steadily rising proportion of carbon dioxide in the air around us may also have affected the climate. Although incoming solar energy can pass through it, carbon dioxide absorbs strongly in the infrared part of the spectrum, where most of the thermal energy that radiates from the ground into space is concentrated. Additional atmospheric quantities of the gas should therefore act as a much improved heat conserver, causing a worldwide warming up. In fact there is evidence that the mean temperature of the Northern Hemisphere did rise by roughly 1.1°F between 1880 and 1940, although we cannot be sure that the rise is directly attributable to the extra atmospheric carbon dioxide. Calculations suggest that a possible doubling of the level from 300 to 600 parts per million during the next century could raise the average surface temperature of the earth by more than 4°F—and that is sufficient to upset the whole delicate heat balance of our planet.

The temperature effect of increased carbon dioxide may be balanced by the presence in the air of dust, grit, and smoke, which leads to a

Smoke from a factory smokestack streams away on the wind. Two centuries of mass manufacturing based on the burning of fossil fuels have greatly changed the content of our atmosphere.

reduction in temperature. Dust also offsets the increased rate of photosynthesis brought about by high levels of carbon dioxide in the air. The dust particles coat the surface of leaves and so the amount of sunlight reaching the chloroplasts is bound to be diminished, thus inhibiting photosynthetic activity. But dust particles are also important in the air, because they can provide nuclei around which drops of water can form. And this is undoubtedly what happens with the dust and grit of domestic and industrial origin as well as with the solid particles sent into the air by high winds, fires, and volcanoes.

As most of us today know, the nature of the air can be changed locally in crowded urban and industrial areas, and occasionally the effluents that are dumped thoughtlessly into the atmosphere are not whisked away by winds but hang around and cause a great deal of unpleasantness. Coal and oil burned on a massive scale produce not only carbon dioxide, grit, and smoke, but also sulfur dioxide, a poisonous gas that reacts with water to form sulfurous and sulfuric acid. In big cities, the atmospheric level of sulfur dioxide is usually around 0.1 parts per million by volume, but it may increase briefly to 1 part per million, when its smell and taste are easily detectable.

Sulfur dioxide may have been the chief culprit in the killer smogs that used to blanket London after the last war. Recent studies indicate that when the concentration of this gas rises above 0.25 parts per million in the presence of smoke, the daily death rate rises correspondingly. But environment-conscious governments in most industrial countries have recently taken measures to minimize the danger of killer smogs. The world's factories still pour out large amounts of sulfur dioxide, but for the sake of local "clean" air, the fumes are expelled through tall chimney stacks, so that the gas is carried down wind for hundreds or perhaps thousands of miles. Somewhere at some time, though, it is precipitated in rain water; it does not just disappear.

Such fumes are also toxic to plants. In many industrial areas, the runoff from the trunks of trees prevents vegetation from colonizing the ground around the trees. Sulfur dioxide from Los Angeles is killing the ponderosa pines in the San Bernardino Mountains, 100 miles down-

Smoke particles in stagnant air produced one of New York City's worst ever smogs in November 1966. Air had become dirtier here than in any other major city in the United States.

wind. The gas in solution is also responsible for tarnishing metal and causing devastating damage to limestone, as the crumbling buildings and statues of Rome and Venice can testify. And women in industrial areas can also blame sulfur dioxide for occasional runs in their nylons.

Another well-known cause of marked local changes in the composition of the air is the automobile. Exhaust fumes contain poisonous carbon monoxide, nitric oxide, and a complex mixture of unburned hydrocarbons. With millions of cars operating in stagnant air conditions, the noxious exhaust fumes can produce very unpleasant smogs. Another source of atmospheric pollution also stems from the car: lead. Tetraethyl lead is an additive in motor fuels and about 75 per cent of it comes out of exhaust pipes and is spread throughout the air and over the ground, much of it in the form of breathable dust.

The combustion of organic fuels high in the atmosphere by aircraft may cause further difficulties down below. Water in jet exhausts produces vapor trails that behave like clouds. These have a high reflectivity, shading the ground or ocean beneath and sending much solar energy back into space. It has been suggested that vapor trails formed on the dense North Atlantic air routes may be reducing the productivity of phytoplankton. The high temperatures and pressures inside airplane engines also generate

nitric oxide. Although the addition of this gas to the atmosphere may be unimportant at the heights flown by subsonic jets, it may well break down the ozone into oxygen at stratospheric altitudes, where supersonic transports are bound to be cruising before long. And a reduction in the stratospheric ozone would weaken the barrier to the sun's damaging ultraviolet radiation.

There is evidence that the atmosphere is becoming dirtier and less transparent by several per cent every decade. Even on the sunniest days in Washington, D.C., approximately 16 per cent less sunlight manages to penetrate to the ground today than half a century ago.

All of us need to breathe clean—or at least harmless—air. The design of our bodies is adapted to an atmosphere composed of the elements described in this chapter. To treat that atmosphere as a waste bin of infinite depth is to gamble with the future of our planet.

Below left: smoky exhaust from this jet airliner over San Diego, California, underlines the problem of airborne pollution. Waste products from low-flying aircraft chiefly affect areas around airports. Below right: privet twigs sicken under a shroud of red dust blown from a nearby steelworks at Consett in northeastern England. In this blighted garden only flowers and leaves newly emerged from underground bulbs are completely free from an airborne pollutant that starves plants of life-giving air and sunlight.

只今の大気汚染状況
亜硫酸ガス濃度
PPM
一酸化炭素濃度
PPM
ここの騒音
ホン
東京都
大気汚染測定場所
亜硫酸ガス濃度：東京都庁前
一酸化炭素濃度：東京都庁前
キヤロン
有馬温泉

Traffic fumes can become a serious problem in some cities, creating unpleasant conditions and health risks to the inhabitants. Left: Tokyo citizens test the efficiency of anti-pollution masks, and (below left) a pollution counter shows the parts per million (PPM) of toxic chemicals present in the city air. Below: smokers' lungs absorb airborne chemicals that can indirectly cause death.

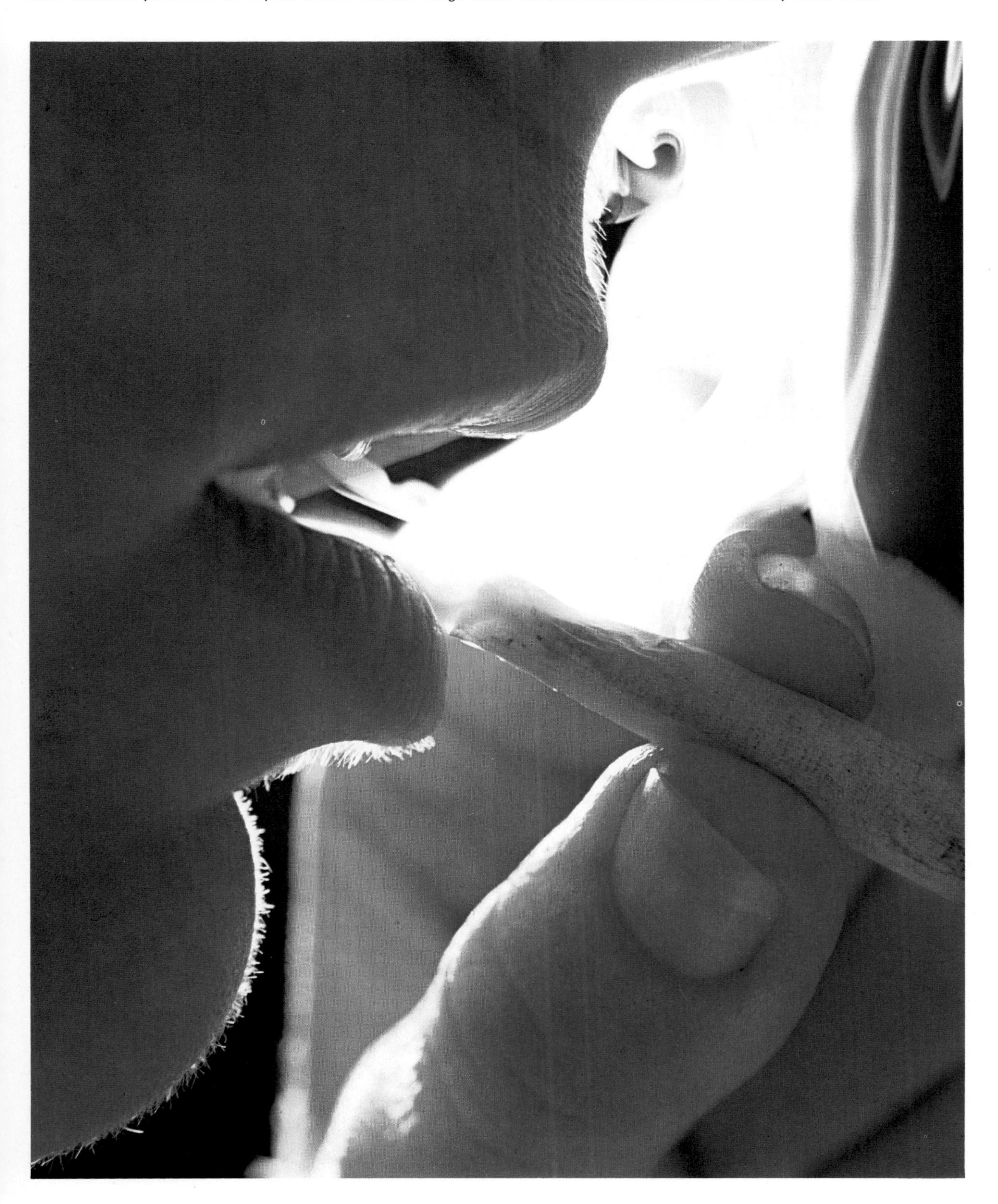

Life in the Open Air

In the beginning everything lived in the open sea. This was where life itself began, and indeed where it remained for around 2000 million years. Conquest of the land came as a very gradual process starting a mere 425 million years ago, at the beginning of the Silurian period. The move from the sea was a major evolutionary advance, because living in the open air posed severe problems that could be solved only by a remodeling of waterborne designs, and the development of complex life-support systems. Because many of the difficulties experienced by animals and plants in emerging from water had less to do with living on land than with survival at the bottom of the atmosphere, the move can rightly be referred to as the "conquest of air."

Water has many advantages over air as a medium to live in. Because water is necessary before such life processes as respiration and photosynthesis can take place it is not surprising that most living things contain more water than anything else. And so what could be better than actually being immersed in it! Water is a fairly dense fluid that supports and cushions the body. Moreover aquatic plants and animals rarely experience violent temperature changes, because water is able to absorb or lose large quantities of heat with little alteration of temperature.

The same cannot be said of air. Over the course of a few hours, the temperature can change from a fiery blast to an icy draft. A land organism is therefore in potential danger of being either cooked alive or frozen to death. The evolution of warm-bloodedness in birds and mammals must be seen as part of a system for temperature control of special value to land animals. In addition, because air is very drying, loss of life-supporting moisture is something that has had to be controlled. And the atmosphere is so thin that it does not give bodily support in the way that water does.

Air seems to offer so few advantages to living organisms, that you may wonder why anything

An ebbing tide leaves seaweed at the ocean's edge. No longer bathed in life-supporting water, the plant lies limply, soon to dry out and die. Once all plants and animals lived in the water: none was designed initially for survival in the open air.

found it profitable to colonize the land. In fact, however, there are some compensations. For instance, there is much more power in the sunlight on land, and a change from the relatively dim subaquatic environment to the brighter surface would enhance photosynthetic activity, and so would hold some attraction for plants. Animals and plants alike also profit from the higher concentration of oxygen in the atmosphere.

Plants must have been the first organisms to come to terms with the difficulties of surviving at the interface between the atmosphere and the land. The earliest settlers may have been rather like algae and were certainly ill equipped to withstand the full rigors of terrestrial life. At first they grew only in very damp places, as liverworts and mosses do today. But they soon developed a system of water retention in the form of a waxy covering to the outside surface, which prevented their internal moisture from evaporating. Water-dwelling algae are bathed in a nutrient solution of salts, oxygen, and carbon dioxide; but on land these resources must be sought in the soil. Accordingly, the organ of attachment, or *holdfast*, of seaweeds gave way to a specialized organ—the root—capable of seeking and soaking up moisture and minerals; and an internal plumbing system evolved, which transported water and salts to the foliage while redistributing the products of photosynthesis away from the leaves.

Leaves also had to adapt to the new habitat, so that they would present as great a surface area as possible to the sun without losing too much moisture. Although covered by a waxy cuticle, one surface became perforated by *stomata*, which are tiny openings that allow an interchange of gases between plant and atmosphere. Numerous spaces within the leaf facilitate the free movement of oxygen and carbon dioxide among the cells. When exposed to strong sunlight, leaves

Below: fronds of fernlike algae thrust upward through water toward the light. Aquatic algae gave rise to the first land plants more than 400 million years ago. Below right: leaflike structures (much enlarged) produced by a liverwort, a moisture-loving plant that colonizes bare damp rocks, as the first land plants may have done. In time, liverworts help to break down rock into soil.

may become very hot; for example, sunlit oak leaves can reach a temperature of 120°F—as hot as a cup of coffee!—when the surrounding air temperature is only in the low 80s. The problem of keeping cool was solved partly by the stomata, through which there is evaporative water loss, and also by the evolution of deeply divided leaves, such as those of oak and ash, which lose heat much faster than do broad blades.

Above all, leaves must have access to light; and in the competitive situation that has always prevailed, land plants were engaged in a race to the open sky. The first colonists were probably limp and flaccid, like their fully aquatic ancestors, which were buoyed up by water. On land the absence of some kind of framework severely limits size, and plants could grow tall only by evolving strengthening fibers. This eventually culminated in the production of long woody trunks and the tree form. The landscape during the Devonian period (405 million to 345 million years ago) was thickly forested, but not with trees of the kind we are used to. They were treelike relatives of the club mosses, ferns, cycads, and horsetails, all soaring upward on massive woody columns to reach a space for their foliage in the canopy

Needless to say, the diversion of so much of an individual plant's resources into making wood is not necessary in an aquatic environment; giant kelps that grow to 150 feet along the Pacific coast of North America are kept afloat simply by a series of gas bladders at the base of their "blades." By contrast, the giant redwoods have the kind of design imposed on any plant that stretches 300 feet into the atmosphere. Any one of their trunks—65 to 100 feet in circumference at the base—contains enough strengthening material to build a small town full of wooden houses.

To survive in the vast Devonian forests, animal life had to surmount some of the same obstacles as the plants. And so the animals evolved their own means of supporting their bulk, conserving water, keeping their temperature within viable limits, and allowing sufficient oxygen to reach their tissues to meet the needs of respiration.

Of the enormous range of animals in the

Silurian and Devonian periods, only a few types were destined to survive on land. The single-celled creatures we call protozoans had no chance of surviving away from moisture; soft-bodied sea anemones, jellyfish, and their kin, built like hollow sacs, needed an ample supply of water for strength and support. Moist-bodied worms managed to colonize the land only by burrowing in the damp soil or taking up residence inside larger animals, where it is warm and wet, thus avoiding all the problems of surviving in the open. Sea urchins, starfish, sponges, and their relatives have never managed to adapt to open-air existence. In fact, only three animal body designs have been flexible enough, in evolutionary terms, to accommodate the changes involved in moving from water into the open air: the arthropods, the mollusks, and the vertebrates.

The arthropods are a major group of animals without backbones, including the chiefly aquatic

Some land plants evolved woody fibers and became strong enough to develop into self-supporting trees. Early trees included ancestors of the low-growing great horsetail (right, showing spore-containing reproductive cones), and the bushy-crowned palmlike cycads of the tropics and subtropics (below).

Yellow gas-filled bladders buoy up the brown blades of this kelp, a kind of seaweed. Thus kept afloat, the giant kelps can grow to lengths of over 150 feet unaided by rigid supports.

crustaceans, the predominantly terrestrial centipedes and millipedes, the arachnids (spiders, scorpions, mites, and ticks), and—most numerous of all—the insects. By any standard, the arthropods have been enormously successful. The first members of the group to appear were the trilobites, which swarmed in the Cambrian seas nearly 600 million years ago, and were perhaps the dominant form of life for tens of millions of years. Like all later arthropods, they had a segmented body, and each segment had a pair of clearly jointed appendages (the word "arthropod" derives from Greek words meaning, roughly, "jointed leg"). Equally characteristic of them and all other arthropods was the fact that the body had a soft center, reinforced on the *outside* by a calcareous crust, rather like a coat of armor. This contrasts markedly with our own design, which is soft and pliable outside, but strengthened by an internal chassis of bones.

We call the arthropods' crust an *exoskeleton*; the supporting chassis is placed on the outer surface, and derives much of its strength from its cylindrical structure. Marine arthropods, such as the trilobites and later crustacean models, hardly needed support, but they did need protection for their tasty bodies, and the exoskeleton gave them this protection. When the first terrestrial insects appeared, perhaps 250 million years ago, they already had both support and protection for their bodies in the form of their exoskeleton.

Although some sort of chassis is necessary, however, there is a limit to the amount of strengthening material that an animal can carry around with it, particularly in the air, which gives no help in the form of buoyancy. In engineering terms the arthropod skeleton is extremely efficient, because it can be regarded as a series of jointed cylinders—and cylinders are strikingly resistant to bending and twisting forces. For its size and weight, therefore, an insect or spider has an extremely strong framework. A look at the detailed architecture of the outside cuticle confirms our impression of strength. Most of this exoskeleton consists of a mixture of proteins and the horny substance called *chitin*. Its most interesting feature is the way the chitin fibers are

Water supports the soft body structures of many sea creatures including the jellyfish (right) and the "grove" of anemone tentacles (far right). Lack of bodily support largely explains why such animals could never live successfully on land.

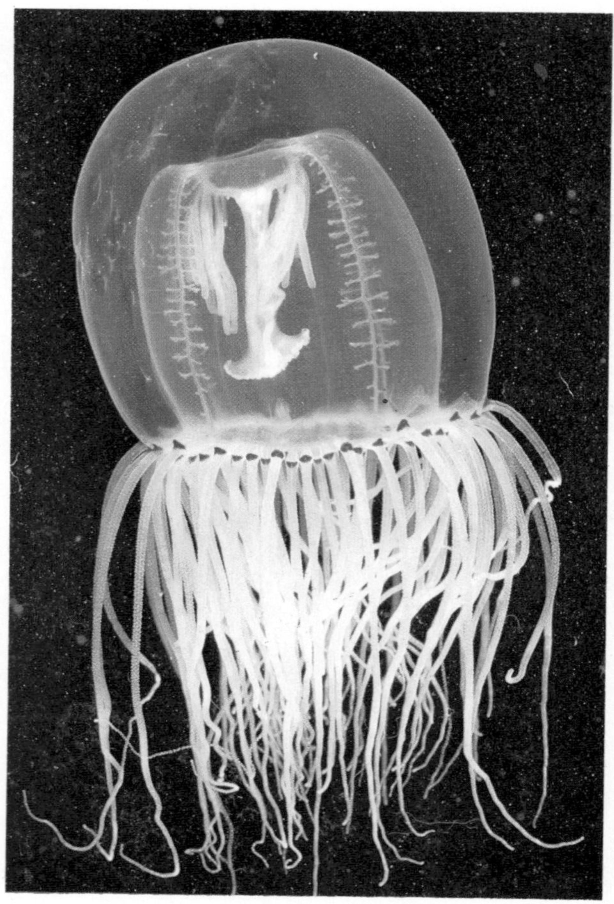

laid down in different directions at various depths, so that the whole cuticle has a structure that resembles plywood or fiberglass. Thus, no matter which way the cuticle is pulled or strained, there are always some chitin fibers placed in such a way as to resist the distortion.

And yet the very feature that gave the arthropod design a head start in surviving away from water has in the long run imposed a severe limitation on the size of these creatures. The cylindrical chassis is their Achilles heel, for the cylinders need to become wider as insects increase in size, and the walls become relatively thinner. Theoretically the cylinders should still retain their stiffness, but in practice wide cylinders with very thin walls buckle and collapse all too easily. This could be prevented by thickening the walls, but it would require far too much building material to be a practical proposition for an insect. Furthermore an exoskeleton is very

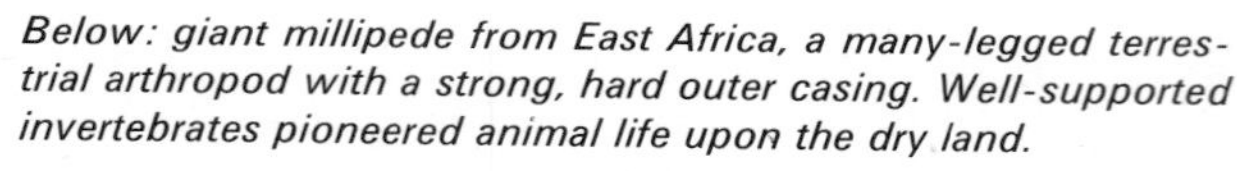

Below: giant millipede from East Africa, a many-legged terrestrial arthropod with a strong, hard outer casing. Well-supported invertebrates pioneered animal life upon the dry land.

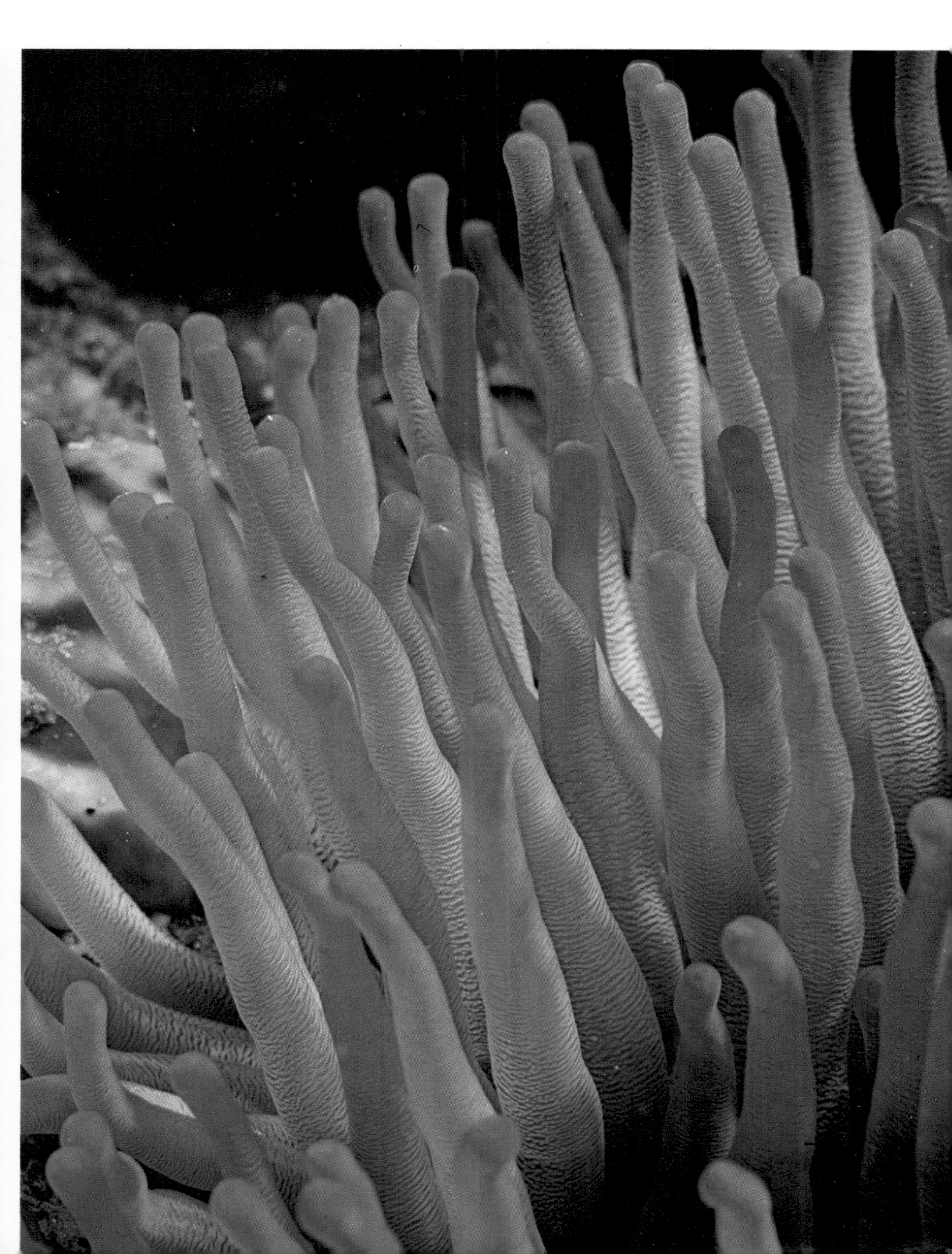

vulnerable to impacts; it can be broken easily, or severely weakened by scratches. Such considerations are not important to small insects, but one the size of a man would literally break up if it stumbled into a tree or fell over.

And there is another disadvantage to the exoskeleton as a means of support. Once secreted by the skin, it hardens and hems the body in. The animal cannot grow without casting off its tough casing. This is just what all arthropods have to do. They molt periodically and grow quickly during the day or two after the old "skin" is cast off, while the new cuticle is still soft and pliable. At such times, the animal is structurally very much weakened, and so it must always be light enough to prevent the newly secreted flexible exoskeleton from becoming deformed during the hardening period. If a land arthropod grew beyond a certain size, it would tend to collapse under its own weight while molting, and the cuticle would harden into an ugly, twisted mass.

Such are the problems associated with living in air. Large marine arthropods have the weight taken off their legs by water. A North Atlantic lobster may weigh over 40 pounds. By contrast, the South American Hercules beetle is probably the biggest land insect; at its largest it weighs less than two ounces! A few land-living crustaceans grow to immense sizes compared with insects. For example, robber crabs, which live largely on islands in the Indian Ocean and the Southwest Pacific, may reach a weight of six pounds. Like all crustaceans, they have strong exoskeletons, but because they molt in the sea they do not have support problems at these times.

Above: male Hercules beetle (life size), the world's longest beetle and probably the largest of all insects. The structural limitations of their external skeletons prevent any land arthropods from growing very much above this size.

Left: this wolf spider's jointed, cylindrical, hard-walled limbs (shown enlarged) illustrate the strong body structure that enables arachnids and other arthropods to move about on dry land.

Right: ghost crab, a type of land crab found on beaches from Long Island south to Brazil. Marine crabs have gills, but crabs adapted to life on land contain big air chambers that act as lungs.

All animals, whether on land or in the sea, must overcome the problem of getting sufficient oxygen into the body, and the problem becomes more difficult with increasing size. This is because, although oxygen is consumed within the bulk of the tissues, it must enter the body across its surface; and as an animal grows, the volume of its respiring tissues increases at a much faster rate than does the surface area. If you double an object's linear measurements, its mass will increase *eight* times, as against only a fourfold increase in surface area. This is why only very small animals such as single-celled protozoans and worms can rely entirely on oxygen diffusing through their skins: they have a relatively large surface in contact with the air in relation to their bulk, and so they can take in enough oxygen to satisfy the body's demands.

Larger animals have overcome the problem by developing highly folded and dissected respiratory surfaces, such as gills, which increase the area of the body over which oxygen can be taken up without significantly adding to the mass of respiring tissue. Aquatic arthropods have gills, but gills do not function well in the atmosphere, for as they dry out they collapse and the surfaces stick together, thus reducing their total surface area. Such arachnids as the king crab "breathe" by means of ventral gill books (so named because the membranes are arranged in folds, like the leaves of a book). Some air-breathing arachnids, such as scorpions, have retained this basic system but the folds have been withdrawn into paired *sacs* or "lungs." Many spiders, too, have paired book lungs. Some spiders show the beginnings of the tracheal system of direct air transportation to the tissues.

Tracheal breathing is characteristic of terrestrial insects. An insect's body is riddled with a maze of pipes called *tracheae*, which open to the outside through a series of breathing holes called *spiracles*. The system takes advantage of the rapid gaseous diffusion of oxygen along these passages to the body cells. The rate is further enhanced by pumping movements made by the abdomen, which also help to maintain a regular

Right: the adult great diving beetle spends most of its life hunting under water, yet it must "breathe" atmospheric air. To get the air it needs, it sticks the rear end of its body up out of the water. The beetle then carries an air supply down under the water on the wings hidden beneath its elytra*—the shiny wing covers on its back. The air trapped under these covers filters directly into the* spiracles*—tiny air holes that supply oxygen to the insect's body.*

Below: prey's-eye-view of a fish's mouth opened to show the gills and gill slits. Water gulped in through the mouth is forced out past the gills—complex surfaces designed to extract the oxygen dissolved in water. All larger animals need special respiratory surfaces, such as gills or lungs, in order to create a sufficiently big oxygen-absorbing area.

flow of fresh air through the system. This tracheal type of breathing has a counterpart in certain plants. For example, reeds with their roots buried in deoxygenated mud have air channels in their stems, so that oxygen can diffuse from the aerial part of the plant down to where the life-giving gas is needed. Swamp cypresses and mangroves also have special aerial "roots," which contain loosely packed cells through which oxygen can diffuse to the deeply buried true root system.

A final major problem that had to be overcome before the arthropods could successfully make the big move from water to land was that small animals living in the open with a large surface area in relation to their bulk are always in danger of drying out. The insects responded to the problem by evolving a remarkable system for economizing water; you realize how effective this is when you consider that some desert species can survive in air temperatures that exceed 140°F, where free-standing water of similar bulk would evaporate in a flash. Their chitinous exoskeleton is a vital weapon in the battle to retain body water, because the outside of the cuticle has a waxy layer that acts as a barrier to the outward passage of moisture. The cuticle also prevents water passing in other than through the correct channel. The comparative lack of success of the crustaceans on land may be due to the absence of a sealing coat of wax.

Passage of water through the body is nevertheless essential for flushing out waste products, particularly nitrogenous wastes from the breakdown of proteins. Aquatic animals take in so much water that they can afford to get rid of wastes dissolved in water. This is not feasible for terrestrial arthropods, which must conserve water. They have overcome this difficulty by changing the chemistry of their waste-disposal system. Instead of excreting highly soluble nitrogenous compounds, which can be ejected only in solution, insects convert nitrogenous wastes to uric acid, which separates out of the solution as white crystals and is then ejected, so that much of the water can be absorbed back into the body.

Most of the water that they do lose goes out through their breathing tubes, and this cannot be helped. After all, a system that permits oxygen and carbon dioxide to pass through it will also let out water vapor. The arthropods have a slight measure of control, however, in that the spiracles can be closed for a while if too much water is escaping through them.

Behavior is another important element in the fight to prevent desiccation. Many small insects (such as springtails) and arachnids (such as mites) actually live in the cool and humid soil or leaf litter, where even moist-bodied skin-breathers can survive. Here are found the wood lice, which are perhaps the most successful of the crustaceans to make the excursion onto land. They manage to survive in a form almost unmodified from that of their marine cousins by avoiding dry, hot surroundings and by reducing their nitrogen metabolism to a bare minimum, so that they have little need to excrete water.

On the face of it, mollusks, which have soft, wet bodies, would not seem to be good candidates for terrestrial life. Yet snails have been prominent members of land communities ever since the Carboniferous period, 280 million years ago. As a group, they have achieved this distinction by a combination of adaptations, tolerance, and behavior. Even when exclusively aquatic, they may have had some features that improved their chances of living on the land. A snail's shell, for example, which supports the delicate visceral mass, offers some protection against both water loss and predators. There is also reason to suppose that many snails emerged onto land from muddy estuaries or marshes, where air breathing already occurred. There would have been little oxygen in the foul marsh water; and so, because of a need to take in air at the surface, the cavity housing the gills—known as the *mantle cavity*—may have been pressed into service as a lung through natural selection. The habit of visiting the surface to take in air is not uncommon among snails today. Some apple snails have the mantle cavity divided into a gill chamber and a lung, so that they can take advantage of both oxygenated water and air.

In fully terrestrial species, the leaflike gills have been lost, and the mantle cavity has been converted into a lung. Many species inhabit very humid places, such as leaf litter or rain forests, and they come into the open air at night, when evaporative water loss is greatly reduced. The snails have also developed the ability of conserving water by excreting uric acid crystals—an

Exposed, stiltlike prop-roots anchor and buttress this red mangrove on a rocky Galápagos island shore. Mangroves often grow in oxygen-deficient mud, where their exposed roots serve a vital purpose by extracting life-giving oxygen from the air.

Some damp-loving worms and snails have evolved special methods of surviving in dry air. Above left: this worm (much enlarged) dug from dry ground had formed itself into a ball to reduce the surface area of its body exposed to the danger of evaporation. Above center: snails cluster on a post and seal off their shell openings to avoid dessication during summer drought in Spain.

interesting example of how certain mollusks and arthropods (and, as we shall see, some vertebrates as well) have evolved similar responses to similar problems.

By no means all snails choose moist habitats; there are some that thrive in dry heat. One species, for instance, clusters on bushes in Africa where midday air temperatures may approach 115°F. Under such conditions they literally shut up home and estivate, sealing themselves in the shell by means of a sheet of mucus, and emerging only when water becomes available. One snail that had adhered to a tablet in the British Museum came "alive" after four years when it was moved into damp air.

The vertebrate body structure differs markedly from both arthropod and mollusk structures. Originally, it may have evolved in the free-swimming larvae of sea squirts. A sea-squirt larva is rather like a tiny tadpole, with a primitive kind of backbone running midway down the body. On either side of this *notochord* is arranged a segmented series of muscle blocks. The flexible notochord acts as a stiffener, so that alternate

Adult sea squirts: fixed, and filtering food particles from water sucked in and squirted out through holes. Larvae similar to those of sea-squirt larvae perhaps gave rise to vertebrates.

contraction of the muscles on each side causes the tail to thrust against the water. In the evolutionary process, this superior design for swimming came into its own among the fish. The notochord became a segmented vertebral column; a skull and jaw apparatus evolved to enable the fish to take in more fuel; and, to aid maneuverability, paired fins developed, which were internally strengthened by means of bones. No other animal has seriously challenged the fish's superiority of design for living in water, and this design was capable of being reorganized to give it supremacy on land as well.

Neither the internal chassis nor the bone-strengthened paired fins bear much weight. In the terrestrial vertebrate skeleton, however, the roles of the vertebral column and the paired limbs have been considerably changed, with the backbone functioning as a girder to take much of the weight of the body, which is largely slung underneath. The heavy load is transferred to the ground by way of the limbs, which have become attached to the spine at its ends by two structures, the pectoral and pelvic girdles.

How was the reorganization of the vertebrate chassis brought about in adapting to life on land? It all started around 400 million years ago, during the early Devonian period when there must have been large areas of stagnant swamps and lakes, which were sometimes flooded and sometimes reduced by the sun's heat to mere puddles. In these conditions, it would have been advantageous to be able to survive being left high and dry and to wriggle across the landscape in search of water. Modern lungfish, which live in habitats that periodically dry up, are among the remnants of a once-large group of fish, to which *Eusthenopteron* belonged. *Eusthenopteron* was a predatory fish whose fins had sturdy lobes. Those on the front paired fins even had an internal framework of bone. Clearly, the fins of *Eusthenopteron* could bear its weight enough to permit the fish to hitch itself along on the ground. But it takes an enormous amount of energy to wriggle forward on nothing but small fin-props. Natural selection must have favored the evolution of limbs long enough to raise the body clear of the ground, and strong enough to lever it forward at speeds beyond the reach of wrigglers. Even back in the Devonian period speed was desirable, because some amphibians were specializing in becoming predators, and others had to keep out of their way.

The lobed fins of *Eusthenopteron* had another vital function to serve: indirectly, they helped the fish to breathe when out of water. The earliest lungfish undoubtedly had one or two sacs at the back of its throat for absorbing air. In order to operate efficiently as lungs, these sacs had to be kept inflated. The danger faced by any aquatic lung-breather when coming onto land is that the weight of its body may cause the sacs to collapse by bearing down on them. This disastrous situation was prevented in the early lunged fishes, such as *Eusthenopteron*, and in their amphibian descendants by the fact that the front lobed fins served as "limbs" to keep the weight off the forepart of the body. Thus the lungs were protected from being squashed, and air could be moved in and out with little difficulty.

Air breathing is not confined to lung fish, but is quite widespread among various other kinds of fish. Those that live in tropical swamps, where the water is permanently deoxygenated by rotting organic material, habitually visit the surface to gulp air. South American armored catfish sometimes make journeys over land in much the same way as *Eusthenopteron* used to. And the mudskippers spend much of their time hunting insects, crabs, and worms on the exposed shore at low tide. There is one curious thing about air breathing, though. It is much easier to absorb oxygen from the air than to expel highly soluble carbon dioxide from the body. So the mudskippers and other amphibious air-breathing fish must return regularly to the water in order to flush the carbon dioxide out through their skin and gills.

As terrestrial vertebrates became larger, they needed to improve the efficiency of their breathing systems. Amphibians with primitive saclike lungs are still basically air gulpers, and part of their oxygen supply finds its way directly into the body through the moist skin. The "hairy" frog of western Africa is so named because of the many hairlike outgrowths of skin along its thighs and flanks, and these are probably a means of providing more surface area over which oxygen can be absorbed. In such creatures, much of the waste carbon dioxide passes out into the atmosphere through the moisture on the surface.

Reptiles dispensed with skin breathing, and all their gaseous interchange with the air takes place in the lungs; these are ventilated by rhythmic expansion of the rib cage, which reduces pressure in the chest and sucks in air. The development of such a suction-pump mechanism means that the major windpipes—the *trachea* and the *bronchi* (which lead from the trachea to the lungs)—must be reinforced with cartilage to keep them from being crushed when the pressure drops. In mammals, the suction pump is further developed by a dome of muscle—the *diaphragm*—that greatly expands the chest cavity when it contracts and flattens.

The mammalian lung is a beautiful piece of biological engineering. Its role is that of a gas exchanger, for it confronts the air with an immense area of moist surface, through which oxygen can diffuse into the blood and from which waste carbon dioxide can be expelled. Our own lungs, for example, have a capacity of between two and three quarts. Nearly one pint is changed with each breath. Because we may take as many as 15,000 breaths a day, our lungs are exposed to well over 1000 gallons of fresh air daily.

Two kinds of fish capable of temporarily breathing air while out of water. Above: the "walking catfish," an African species introduced to Florida. Air chambers over its gills allow this fish to forage overland (at night or in damp weather) using its fins as legs. Right: the mudskipper (shown here considerably enlarged) breathes out of water through its gills, which are kept moist by the tightly closed gill covers, or opercula. *It may also absorb oxygen through its tail, which is often kept submerged.*

The air is passed down a series of reinforced tubes—the bronchial tree—that branch irregularly, and terminate ultimately within the lungs in a vast series of delicate blind sacs (the *alveoli*). The walls of each alveolus are extremely thin, and are surrounded by a dense network of capillaries through which blood flows. The alveoli, of course, are a means of bringing the air and sheets of blood close together.

In a human adult there are approximately 300 million alveoli, adding up to a respiratory area of about 85 square yards. (For comparison, the surface area of the skin is likely to be no more than 2 square yards.) Before the oxygen can pass into the blood, however, it must dissolve into the film of moisture that coats the alveolar membrane. Once in solution, it is soaked up by the hemoglobin in the red blood corpuscles; at any one moment an estimated 2300 of these corpuscles are in the capillaries surrounding each alveolus. The amount of oxygen that passes from the air across the surface of a human lung varies according to the activity, the depth of breathing, the amount of hemoglobin in the blood, and the rate at which the heart pumps blood through the arteries. At maximum effort, the lungs may be able to absorb 10 pints of oxygen in one minute, but normal consumption is very much less.

Mammalian and reptilian gas exchangers have one drawback. Because they are based on a tidal flow of air, the air inside the lungs cannot be completely changed with each breath. The birds have developed a one-way system of ventilation that—in theory, at least—allows for a much

more efficient utilization of incoming air. Whereas the mammalian system has so many dead endings that the air cannot flow freely, the birds' system has no dead space where fresh air becomes mixed with stale gas. The oxygen requirement of birds is very high, of course, and the architecture of their breathing system may be related to the high demand. The respiratory surface of a bird lung is about 10 times larger in relation to body surface than that of a mammal. Throughout the bird's body there are membranous air sacs that prevent the lungs becoming clogged up with stale air. Air from which the oxygen has been removed is not forced back the way it entered the lung, but is shunted into the sacs to make room for fresh air. Expired air comes from these sacs, preserving the one-way flow through the lungs.

Like insects and other invertebrates that live in the atmosphere, vertebrates must also solve the problem of conserving water. Air, which is warm and saturated in the lungs, takes a lot of moisture with it when it is expelled, as you know if you have noticed your steamy breath on a cool day. About 15 per cent of our total daily water loss is accounted for by its disappearance through the mouth and nose.

Many mammals and reptiles—although not man—have evolved a system of air conditioning that reduces the amount of water exhaled. The relatively dry air that the animal inhales meets a number of moist baffles in the nasal passage. Water evaporated from the nasal surfaces cools down both the air and the nose. And so when hot saturated air from the lungs meets the cool nasal surfaces, its temperature is reduced enough to allow some of the water to condense out within the nose. Enough moisture is saved by this cold-nose technique to make a significant contribution to the water budget of such animals. In the desert iguana, for instance, cooling of the exhaled air reduces the respiratory water loss by over 30 per cent in comparison with what it would be if the iguana had to expel the air at lung temperature.

Water is also saved by the excretion of uric acid crystals by the kidneys of many reptiles and birds. In man and other mammals, an efficient kidney permits the formation of a very concentrated urine by reclaiming most of the valuable sugars, salts, and water from dilute waste and leaving chiefly such substances as urea and surplus water to be excreted. In their daily business of purging the wastes from our blood, the kidneys filter out more than 30 gallons of fluid, but most of the water is absorbed back into the blood, leaving perhaps 3 pints of concentrated urine that must be voided.

The other major way in which vertebrates conserve moisture is through the waterproofing of the skin. Amphibians and fish out of water lose a great deal of moisture through the secretion of mucus to keep the skin damp. Early in the evolution of terrestrial life, a damp skin not only helped animals to breathe but also kept their temperature down by evaporative cooling. This was a wasteful practice, satisfactory only as long

A puffin in flight. Birds need a big oxygen intake. To meet this need, they have specially designed breathing systems. Air sacs throughout the body improve breathing efficiency.

as there was a plentiful supply of water. Eventually the reptiles, birds, and mammals became less reliant on water as they evolved a dry, relatively waterproof skin, with a highly efficient mechanism for holding down body temperature.

One disadvantage of living on the land is that it is much harder to keep cool in air than in water. To cope with this problem, many birds and mammals get rid of their excess heat by evaporating water from the body surface. When a dog pants, much of its heat is lost from the moist surfaces of its nose, tongue, and mouth. Like some other mammals, we do a similar job of keeping cool, but by means of sweat glands—between 2 and 3 million of them per person.

Although scattered throughout the skin, the sweat glands are densest on the palms of the hands and on the soles of the feet. Each gland is a coiled tube, surrounded by a network of blood vessels, which opens onto the surface. An

Exhaled moisture condensing inside the desert iguana's nose helps to conserve water and thus to prevent this southwestern lizard drying up in the air heated to a temperature above 100°F.

individual's output of sweat may be one or two pints a day, but under conditions of extreme heat the skin may lose as much as one gallon. The fact that a person can sit for quite a while in a sauna "bath" at 230°F—well above boiling point—providing the air is kept very dry to promote evaporation of sweat, testifies to the effectiveness of this system of keeping cool.

The skin of birds and mammals plays a very important role in regulating body heat. This is because air is a poor conductor of heat; as long as a stagnant layer of air can be held close to the body, heat can be prevented from flowing in or out, to a greater or lesser extent, as the situation warrants. The thicker the boundary layer, the more effective it becomes as an insulator. Fur and feathers, which are part of the skin structure, make use of this fact. The winter pelts of such cold-climate mammals as Arctic foxes have extremely thick fur as compared with that of similar hot-climate animals (for example, the African bat-eared fox). Mammals that live in tropical areas tend to have coarse hair that gives protection from the strong sunlight while allowing cooling air to circulate freely around the skin. And many birds control the thickness of air trapped in their plumage by either flattening the feathers to cool off or fluffing them in order to keep warm.

During the course of evolution, human beings have replaced thick fur coats with layers of fat beneath the surface of the skin. What little body hair we have retained may seem of small consequence. Yet it helps to keep a thin layer of air clinging to the skin. This barely noticeable air

The walrus is insulated from winter cold by a thick skin underneath which is a layer of fat, or blubber some 2½ inches thick. In summer, the walrus radiates excess heat through blood vessels in the skin, which become dilated and turn the walrus a rosy color.

Three ways in which different land mammals avoid overheating. The African elephant (left) has huge ears that serve as fans and radiators to dissipate surplus body heat. The hippopotamus of tropical Africa (above) lacks such built-in radiators and cannot keep its vast bulk cool in hot air. Instead, it escapes the fierce heat of the sun by wallowing in mud or water. The spaniel (right) keeps cool by panting—a process that produces heat loss by evaporation. Heat drawn from the dog's tongue, mouth, and nose helps to transform the moisture present on those surfaces into water vapor.

coat shields us to some extent from violent extremes of temperature. For example, if you break down your own boundary layer by blowing hard on your naked arm while sitting in a hot sauna, the skin will quickly blister!

Increasing size imposes extra burdens on warm-blooded mammals and birds, because air is such a poor heat conductor. After all, the production of animal heat is proportional to bulk or the volume of tissues, but loss of heat occurs all over the body's surface. It follows that little creatures with a small amount of heat-producing tissue, but with a relatively large surface area, will probably have difficulty in maintaining a steady body temperature. They tend to have a high metabolic rate and thus a high rate of heat production, and they also need thick pelts, which enable them to retain their body heat.

Large mammals on the other hand, have plenty of bulk but relatively little surface over which to lose their excess heat. It should come as no surprise, then, that big mammals living in hot

climates have either sparse coats or practically naked bodies, to facilitate heat loss. African bush elephants have ear flaps that do double duty as either fans or radiators. When held clear of the body, such flaps increase the elephant's skin surface by one sixth, or even more. Similarly, the ostrich dissipates heat from its naked thighs by a fanning motion of the wings, whose plumes of fluffy feathers are excellent for creating a draft. It is even possible to believe that the "sails" that grew on the backs of the pelycosaurian and spinosaurian reptiles that flourished between 100 and 200 million years ago were useful for adjusting the body heat by either absorbing warmth from the sun or radiating it.

The change of life style required by a move away from water to land involved fundamental modifications not only to the body but also to breeding methods. Sexual reproduction is relatively easy in a watery medium. Sperm and eggs can be liberated directly into the water, where they can mix and meet by chance. Many aquatic animals, such as sea urchins and herrings, reproduce in this rather hit-or-miss fashion. But any such system of external fertilization is impracticable on land, where the warm air would very quickly dry up seminal fluid, and where ejected eggs could be protected against desiccation only by means of casings that would not admit sperm.

The much less chancy method of internal fertilization—the injection of sperm into a female while the eggs are still inside her—was probably first practiced by the mollusks, particularly the snails. Many of the more advanced gastropods, having developed an appendage for squirting sperm directly into the oviduct of a mating partner, were thus pre-adapted to breeding in the atmosphere, and they were undoubtedly among the first animals to make themselves at home out of water. Terrestrial snails lay yolky eggs with protective coatings to cut down water loss, and the larval stage takes place inside the egg.

Internal fertilization takes place, too, in many crustaceans, such as isopods, which have also dispensed with free-swimming planktonic larvae. Instead, the female retains the eggs and young in her body until they are able to look after themselves—a primitive kind of parental protection that obviously pre-adapted the isopods to a precarious open-air existence. The arthropods also evolved various approaches to terrestrial sperm transference. Scorpions, for instance, keep the wet seminal fluid away from the drying effects of air by packing it in a sperm parcel (called a *spermatophore*), which the male deposits before maneuvering the female into a position where the vital package is received into her genital aperture.

Reptiles were the first vertebrates to dispense with gill-breathing larvae and the need to find free-standing water in order to breed. Male and female reptiles copulate; the male pours sperm into the female's cloaca, and fertilization takes place high up in the oviduct. Perhaps the rep tiles' most remarkable adaptation to survival in the atmosphere occurs in the egg itself. With no water to support them, large soft eggs would collapse when laid, and so a tough leathery or calcareous shell builds up around the fluid contents as the egg passes down the oviduct. The shell protects the soft center against the action of gravity, and also helps to keep water loss to a minimum. Inside a reptile egg, the delicate embryo is enclosed in a thin membrane containing what is called *amniotic fluid*. Thus the embryo has its own private pond, which prevents it from drying up.

A further membrane lies outside the amniotic membrane, close beneath the surface of the shell, and functions as a primitive lung. This membrane takes oxygen in blood vessels to the developing reptile, and also acts as a kind of refuse bin for wastes. Such systems were essential to the development of land vertebrates, and they exist in birds and mammals as well as in reptiles.

Once a reptile's eggs are laid, they are usually warmed by the heat of the sun. But whether they hatch may depend on the weather, for they can be scorched and desiccated, or cooled by rain, or flooded and cracked. It is not surprising, then, that some snakes and lizards bear their young alive. In many cases, the eggs are simply retained inside the female's body to the point of hatching—which does, of course, protect them from the elements. But a number of relationships between the parent and her young are more prolonged, and therefore even more protective. For instance, some skinks (a kind of lizard) nourish their offspring by means of a placenta.

Higher mammals have improved on this system of internal fertilization and retention of eggs. Following copulation, the fertilized egg quickly establishes contact with the uterine wall, and it thereafter gets all its nourishment from the mother's blood supply by means of a placenta. Many mammals, such as horses and other ungulates, are even born in a sufficiently advanced state to cope with most of the problems posed by living in the open. For instance, they can stand

Spiny lizard with young placed beside her for a size comparison. Some kinds of spiny lizard produce liveborn young while other species of this North American group of reptiles lay eggs.

almost immediately and can run in a few hours.

But internal fertilization, although the most practical way for animal species to ensure their survival in open, dry places, does bring with it some special problems. The sexes must meet, recognize each other, coordinate their moods, and stay in contact long enough to permit transference of seminal fluid from male to female. Courtship is an answer to many of these problems, because it is a language that opens up the path to mating.

The form the language takes varies from one species to another. Some male spiders, in constant danger of being mistaken for a meal by their larger and more ferocious mates, go courting with a fly wrapped in silk; the male thrusts this into the female's jaws as he gets down to the job of filling her up with sperm. Visual signals have been adopted by other spiders, as well as by birds and some mammals. And there are exotic perfume languages conveyed through the atmosphere. Female moths, for example, produce seductive odors from glands at the tips of their abdomens when they are ready to mate, and these may attract males from miles around. Airborne sounds are also commonly used in courtship by insects and birds.

Plants have also had to evolve a sex life that does not rely upon water. The way in which they came to terms with dry and airy places can be surmised by looking at existing types of plant. To complete their life cycle, such primitive kinds of plant as mosses and ferns pass through separate asexual and sexual stages. The mosses and ferns that we all recognize as such are asexual *sporophytes* (spore producers). They liberate millions of spores into the air, and a few of these fall on damp soil and grow. But each spore becomes not another moss or fern plant, but an intermediate form called a *gametophyte*. This is an extremely small but nevertheless important stage, because each gametophyte produces both male and female sex cells.

The eggs, or *oocytes,* are usually retained in

Below: baby giraffe during birth (left) and being cleaned by its mother shortly afterward (right). Internal fertilization and development make it possible for mammal species to survive in open, dry places. Also giraffes are so well-developed at birth that they can stand and run only hours after being born.

flasklike structures, whereas the sperm are mobile and swim about in surface moisture until they meet an oocyte. Once fertilization has taken place, the oocyte grows into the familiar spore-forming plant again. Obviously, any plant with this method of reproduction is reliant upon water for the completion of its life cycle. Such plants as mosses, ferns, and liverworts therefore tend to grow where the ground is wet.

The complete breakthrough for dry-land plants happened in two stages. First came the production of two kinds of spore, one germinating to form a female gametophyte, the other a male gametophyte. Secondly, the independent gametophytes were dispensed with and the sex cells were retained within the spores themselves. This was eventually followed by a system of reproduction that is in many respects parallel to that of land animals, where the egg stays at home and the male must, so to speak, come knocking at the door. In other words, the female "spore," enclosed in the plant's ovary, awaits fertilization by the male pollen grain. As there is no longer an independent, free-living gametophyte to bring the pollen grain within easy reach of an egg, the problem faced by higher plants is how to get them together. Birds and insects do much of the job of transferring pollen from flower to flower,

Left: a pair of albatrosses courting in the Falkland Islands. Visual signals help many birds to coordinate their moods and come together for the mating process—the biological object of which is to enable sperm to meet egg in the moist environment inside the hen bird's body. Only under such conditions can fertilization of eggs occur among nonaquatic animals.

Below: ringtailed lemurs from Madagascar are among the many animals in which airborne scents play an important role in the preliminaries that lead to mating. In the mating season, a male lemur transfers scent from wrist glands to his tail, then flicks the tail above his head. By this action he visually asserts himself and at the same time spreads his scent upon the wind.

but the air itself is also an important agent.

Among the many plants that scatter their pollen to the winds, grains, cereals, conifers, and catkin-bearing trees send their male sex cells forth in such profusion that they sometimes create clouds of yellow dust. In pines, the pollen grains are made buoyant by air-filled bladders. The chances for pollen to make a fruitful landing have been further increased by the evolution on top of the ovaries of feathery plumes, or highly receptive stigmas, which help to trap airborne pollen. Plants that are pollinated by windborne pollen are called *anemophilous*, and the flowers of anemophilous plants are sometimes raised on stalks, so as to be accessible. In other such plants, flowering occurs before the leaves begin to appear; thus there are fewer hazards in the way of successful fertilization.

Survival in the atmosphere, then, required changes in the breeding techniques of all forms of life. But still further physiological changes were necessary for the animals: their sense organs had to be profoundly modified, because sensory equipment adapted to working under water is next to useless in the air. The vertebrate eye and ear are particularly good examples of redesigning to cope with life on land.

Fishes' sense organs are tuned for the con-

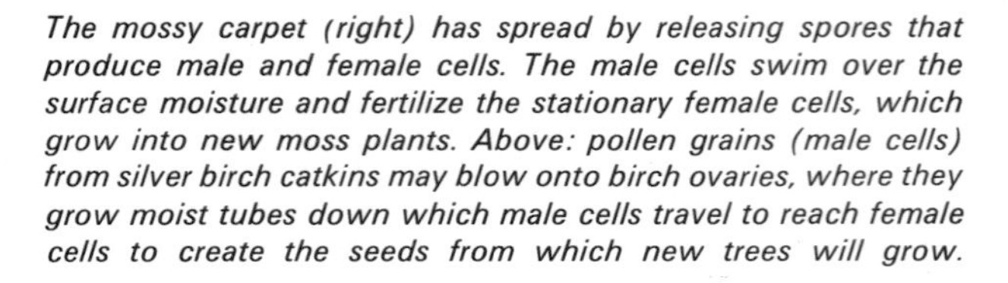

The mossy carpet (right) has spread by releasing spores that produce male and female cells. The male cells swim over the surface moisture and fertilize the stationary female cells, which grow into new moss plants. Above: pollen grains (male cells) from silver birch catkins may blow onto birch ovaries, where they grow moist tubes down which male cells travel to reach female cells to create the seeds from which new trees will grow.

ditions that normally prevail beneath the surface, where the water is often turbid and the light is relatively dim. As light penetrates deeper, the red and yellow wavelengths become increasingly absorbed, so that even at a depth of 95 or 100 feet, everything is bathed in blue-green light. The fish's vision, therefore, tends to be very sensitive to the blue-green part of the spectrum, and its eyes have a relatively large transparent window (the *cornea*) to let in as much light as possible. Because it is also difficult to focus an underwater image on the back of the retina, a fish possesses an oversized spherical lens to deflect the light passing through it.

On land, the light is intense and the color spectrum broad. And so day-active terrestrial vertebrates have evolved relatively smaller corneas; and the curved cornea itself has taken over a major share of focusing the light, with only fine adjustments left for the much flatter lens. Moreover, the sensitivity of the eye is shifted rather more toward the yellow part of the spectrum than to the blue-green. In short, because air is more transparent than water, the physical act of seeing is easier in some respects for terrestrial than for marine creatures.

Hearing, however, is a different story. The transmission of vibrations—or sounds as we perceive them—is more difficult in the atmosphere than in water. The reason why fish have no external ear is that they need none. A fish's body has approximately the same density as the water, and so sound waves pass right through the body tissues and reach the inner ear with undiminished intensity. On the other hand, air has a very low density. Airborne sound waves have so little energy that they are blocked by the dense body of the listener, and there must be a channel through which they can pass to the hearing system and a device to amplify the vibrations.

The mammals and birds have perfected the ear as a mechanism for hearing. In mammals, a delicate drum is stretched across the ear canal, so that any pressure waves passing down it cause the drum to vibrate. We hear sounds when the pressure waves hit our eardrums at between 15 and 15,000 times per second. Of course, the atmospheric pressure itself is always changing, and the design of the ear incorporates a passage —the *Eustachian tube*—that connects the middle ear to the throat, so as to equalize pressure on both sides of the eardrum. Failure to equalize the pressure would result in an over-taut drum and partial deafness. We have all experienced this disability when a rapid change in air pressure occurs in an elevator or an airplane.

The senses of sight and hearing not only provide air-dwelling animals with windows to the world outside themselves, but serve as receivers for messages transmitted by other animals. Because the atmosphere is so transparent, airborne visual signals carry a long way in open habitats. Admittedly, some fish have evolved a language of colorful signs, but these tend to operate over only small distances and in habitats where the water is clear enough for effective transmission (for instance, around coral reefs). There is no underwater sign language, though, that compares with the extraordinary eye-catching courtship display of the cock great bustard or the ostrich, whose posturings can be seen a long way off in the open habitats where they live.

Sound production is much more useful for signaling in such habitats as woodlands or reed beds, because, unlike airborne visual signals, sound travels around corners. Some animal species even produce instrumental sounds; deathwatch beetles knock their heads against their burrows, many woodpeckers beat out

Because light is brighter in air than under water, the eyes of land vertebrates need relatively smaller corneas ("windows") than the eyes of fish possess. You can get some idea of the difference by comparing visible eye areas for this orang-utan (left) and the giant sea bass (right).

tattoos against resonant dead branches, and chimpanzees thump on tree trunks. Howler monkeys, gibbons, birds, and insects have all independently developed sound languages for long-distance signaling. Insects, of course, have no "voice," but many of them can set the air in motion by rubbing parts of their body together. The noise produced by such *stridulation* (to give it its proper name) is generally amplified by resonating wings or wing cases. Cicadas have a powerful mechanism in the form of a pair of timbals sunk into the abdomen, and the diaphragms of these are clicked in and out by muscles.

It is all too easy for us to be unaware of the sophisticated control mechanisms and engineering that living creatures have evolved in order to survive at the bottom of the atmospheric ocean. The long evolutionary journey that we and the animals and plants that surround us have taken has proved to be something of an obstacle race. We are able to perceive the riot of spring colors, to sing and talk, to sunbathe without roasting, and to enjoy the pleasures of making love—and these are only a few of the consequences of our ancestors' having cleared the hurdles on their way from the sea to the open air.

The Rain and the Wind

For the past 3000 million years there has been a close relationship between the development of the atmosphere and the evolution of life on earth. Climate and weather, which are mainly created by the air around us, still profoundly affect the lives and distribution of animals and plants. Climate can be a dominating influence on the character of landscapes. For example, warmth and wetness all the year round allow the growth of tropical jungles, which are natural treasure troves, with an incredible diversity of species. Conversely, cold, windswept areas can muster only a scattered selection of living things.

The daily weather patterns that add up in the long term to the climate are caused by great masses of air rising, falling, swirling, and mixing, for the atmosphere is never still. Some of the motion is due to the fact that the envelope of gases rests on a spinning globe; because air is thin and not very sticky, it is not dragged along at the same speed as the earth, but tends to lag behind. A more important source of turbulence, however, is the sun. Its powerful rays warm the earth's surface, which in turn heats up the air. As the air is warmed it expands and so decreases in density. This hot air rises—a fact made use of by balloonists, glider pilots, and soaring birds—and creates a region of comparatively low pressure, which draws in cooler air at ground level. This action of rising, sun-warmed air provides the basic motive force behind atmospheric circulation.

The sun does not strike with equal power over the whole world, however. Land heats up much more quickly than water; and where the sun's rays slant at a shallow angle, their energy is sapped by the long passage through the atmosphere. For this reason, the polar summer sun has only a fraction of the warming ability of the directly overhead equatorial one. All the many aspects of such inequality in heating set up the massive upward and downward movements of portions of the atmosphere that meteorologists

This lone tree has been forced to grow in a grotesque, lopsided manner by strong winds from the sea, blowing right to left as we look at this New Zealand scene. Climate and weather deeply affect the lives of plants and animals everywhere.

Only the inner sun is real. Outer "suns" are light rays bent by ice crystals in cold air. The sun's rays lose much heat as they slant down through the atmosphere to the polar regions.

speak of as *convection currents*. And in very complex ways, the air movements are further affected by forces generated by the earth's rotation: the winds blow—or, rather, whirl and eddy—from areas of high to low pressure in an attempt to even out the pressure differences.

Broadly speaking, convection currents play an important role in redistributing solar heat more equitably between different areas. Temperature is, of course, critical for the well-being of living organisms, and, luckily for us, we are relatively snug in our immense spaceship, which orbits neither too close to nor too far away from our hot star. If the earth received about 10 per cent less solar energy, its temperature would plummet and the oceans would turn to solid ice. An appreciable increase in the power of the sun would roast the world, and all of its water would vanish into the atmosphere as scalding steam. As it is, earth's surface tem-

Lenticular (lens-shaped) cloud near Nelson, New Zealand. Where clouds form, atmospheric moisture yields heat. Clouds and wind together shift heat poleward from the equator.

perature varies from about −60°F to +120°F, quite a small range, by astronomical standards. The average world temperature hovers around 60°F, and it is this comfortable level of heat that has allowed the evolution of life on our planet.

Life as we know it is based upon carbonaceous molecules linked together in complicated patterns in water. This molecular architecture, with few exceptions, can exist only within a comparatively narrow temperature range—between 32°F and 120°F. Below 32°F water solidifies and is no good as a solvent, and above 120°F the protein molecules coagulate. Our world offers these conditions over much of its surface, and it also has plenty of water. Without water, there would not only be no life, there would be little weather, for weather is, among other things, the daily coming and going of clouds, the alternation between rain and drought. And the transporting of water vapor in the atmosphere plays a crucial role in maintaining much of the earth's surface at a comfortable temperature and keeping it habitable.

Our world is peculiarly blessed with an abundant supply of water. Two thirds of the earth's surface is covered by it, and some 97 per cent of this is held in the sea, with the remainder stored as ice, ground water, and atmospheric vapor. The water is constantly circulating. Each year a monumental 95,000 cubic miles of it evaporates into the air—about 80,000 from the surface of the sea and 15,000 from the land and the plants that grow on it. For example, it has been calculated that an acre of corn will give to the atmosphere 1300 tons of water during the growing season.

Clouds are water in transition. Even a small billowing cumulus may contain as much as 1000 tons, sucked up from the land or sea. About 95 per cent of clouds evaporate away; the rest, which drop their cargo of water as rain, hail, sleet, or snow, provide the world with an average 40 inches of annual precipitation—enough to give every human being over 20,000 gallons a day. Although, of course, it does not fall conveniently where it is needed, the amount condensing out of the atmosphere does balance the water lost from the surface by evaporation. At any one time the air itself contains only a tiny 0.001 per cent of the world's water. Yet the constant turnover

Above: torrential downpour in a part of the Zaire (Congo) Basin, which has hot, moisture-laden air and year-round rains. Left: an ocean wave. Water evaporating from the sea provides 84 per cent of all atmospheric moisture. Hence the sea supplies the majority of the fresh water that life on land needs for survival.

between this minute proportion of vapor and free-standing water plays a vital role in moderating the climate of our planet.

The physical properties of water make it ideal as a heat carrier, operating through the sun-powered water cycle—from earth to atmosphere and back to earth again. As water changes its state from ice to liquid to vapor, it absorbs a great deal of energy in the form of heat. Think how much power it takes to boil a kettle dry; the power is needed in order to energize the water molecules sufficiently so that they are shaken apart and become gaseous. This so-called *latent heat* of evaporation does not vanish into thin air with the steam, but is stored in the vapor molecules themselves. Then, when the vapor condenses out to form water droplets, the latent heat reappears and warms up the air. Similarly, when water crystallizes into ice, a further instalment of latent heat emerges. This is why refrigerators have to work so hard to make ice cubes: the cooling system must remove a great deal of heat before the molecules are deenergized enough to settle down into the rigid structure of ice. Water vapor and water droplets are therefore a highly effective means of transporting and redistributing the heat of the sun from warm to cool areas.

The pattern of heat flow follows the general circulation of the atmosphere. In the tropics huge quantities of heat are absorbed by the evaporation of vast amounts of water. The vapor is carried up and away on hot convection currents, and the air expands and slowly cools as it rises. Cool air cannot carry as much water vapor as warm air, and so the cooling causes some of the load to condense, and there to give up its latent heat. A transport of heat toward the

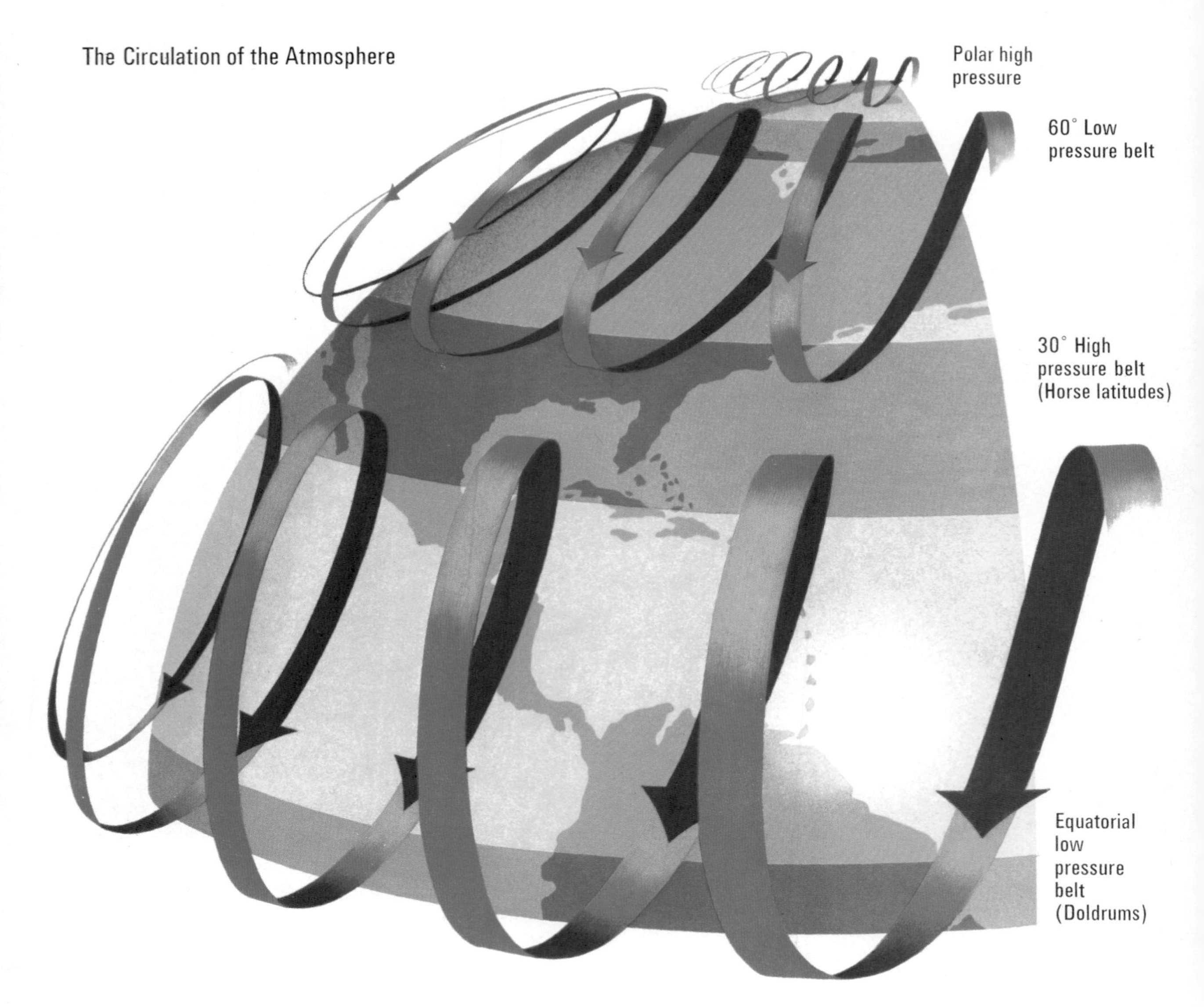

Arrowed loops represent the Northern Hemisphere's three atmospheric cells, which transfer heat and moisture from equatorial regions to the north polar area. The equatorial cell features warm air rising near the equator. As the air rises it thins out to leave a low-pressure belt at low altitudes. As the air cools, it sinks and "thickens," to create high pressure at low altitudes in the horse latitudes. The polar cell works roughly similarly. Between polar and equatorial cells lies the mid-latitude cell, its circulation deeply influenced by jet-streams—high-level, high-speed winds that whirl eastward around the globe.

poles is therefore brought about through the agency of rain. When moist air reaches the poles, it is cooled further, and the water it contains forms ice crystals, liberating the heat picked up by water vapor perhaps thousands of miles closer to the equator. This does not turn the poles into subtropical hot houses because much of the heat is liberated so high that it radiates out into space. But the imported heat does greatly modify the frigid temperatures.

The climate varies with latitude, and the way it changes from the equator poleward is determined partly by the circulation of the atmosphere, which produces fairly well-defined wet and dry zones that girdle the planet. Temperature and rainfall, or lack of it, work together with other factors to mold the kinds of life that inhabit the various zones. Let us begin our examination of how weather affects the living environment at the equator—where, it could be said, all weather starts.

The intense heat of the sun creates a massive low-pressure area around the equator. The hot air expands and rises, and there is very little

Forest giants crowd the slopes of the Crocker Range in north Borneo. These trees form part of the world's great equatorial forest belt, produced by a hot, wet climate throughout the year.

horizontal movement or wind. Seafarers used to call these equatorial calms the "doldrums" because the absence of any kind of breeze would keep their sails slack for days on end. The persistent low pressure over the land draws warm, damp air from the sea. Surging skyward, this air, heavy-laden with moisture, expands, cools, and jettisons its cargo of water. Accordingly, the equatorial climate is characterized by cloudy skies and almost daily torrential rains.

The hot, wet climate produces a belt of lush rain forests within 4° latitude on either side of the equator—a belt extending from the Amazon basin through Africa, Burma, Thailand, and the Malay Peninsula to the islands of Sumatra, Borneo, and New Guinea. Included in the zone are some of the wettest places in the world, for the forests here are drenched with an average annual rainfall of 160 inches, with nearly three times as much as that in a few places. The climate is very stable because the sun is always high in the sky here, the rainfall is spread fairly evenly throughout the year, and the temperature stands at a constant 70—85°F. Even at night the clouds prevent a great deal of the daytime heat from escaping, and so it remains warm, with an oppressive humidity of almost 100 per cent.

The warm, damp, steady conditions generated in the equatorial low-pressure systems may be responsible for the tremendous diversity of life in the rain forests, where one acre is likely to contain as many as 60 different kinds of tree. In Southeast Asia alone, the forests support as many as 1500 species of vertebrate and 150,000 kinds of invertebrate. With no marked seasonal change, the trees have their own rhythms of flowering and shedding their leaves, so that there is always food somewhere for herbivorous animals. And the humidity particularly suits small invertebrates. Even normally aquatic leeches and flatworms can survive in the open, for the saturated air allows very little water to evaporate from their moist bodies.

For the same reason, the rain forests are a haven for frogs and toads. The daily drenching provides all the water they need for breeding. Some lay their eggs in the scroll-shaped rain-water-catching leaves of such epiphytes as the bromeliads. Others, such as *Chiromantis rufescens,* a small leaf frog from the African rain forests, make nests of foam for their eggs. The female has a foam gland in her oviduct, and while she is egg-laying, she whips the secretion from the gland into a froth. This hardens on the surface, but softens before the tadpoles emerge. Such a system of reproduction depends for its success upon near-saturated air, for the foam nest would dry up in a less humid climate.

Because the earth is canted over on its axis, the Southern Hemisphere tilts toward the sun for half of the year, and the Northern Hemisphere for the other half. As a result, the boundaries of the tropical low-pressure system, with its attendant rain, move steadily north of the equator after mid-March and south after early October. Between latitudes 4° and 10° North and South, in places such as Guiana, there are two distinct wet seasons, whereas in the zone lying between latitudes 10° and 23° North and South only one distinct period of rainfall occurs every year. We call this the *monsoon.*

In January, while the sun is beating down on the Tropic of Capricorn, at latitude 23½°s, the earth's heat equator is at its most southerly limit; sucking in warm air from the Southern Atlantic, Indian, and Pacific oceans, it deposits torrents of rain over a belt from southern Brazil, through Natal, to Southeast Asia and northern Australia. In June the situation is reversed, and the monsoon rains fall north, around the Tropic of Cancer. The great land mass of Asia has a very marked influence at this time of year, for it is tilted toward the sun, and the solar heating helps to create a deep low-pressure center over Pakistan and northwestern India. This, together with changes in circulation at higher levels, causes great volumes of wet air from the Indian Ocean and the Bay of Bengal to flow north and northwestward and to fall on the Indian peninsula as the summer monsoon.

Once started, monsoon rains can pour down for months at a time, turning the scorched land into a quagmire. Cherapunji in India, an outstanding example, gets nearly 440 inches (over 36 feet) of rain every year. It is dumped on the area during the single rainy season!

As winter approaches, the land begins to cool, and the low-pressure system is replaced by cooler, descending, dry air. At first, while

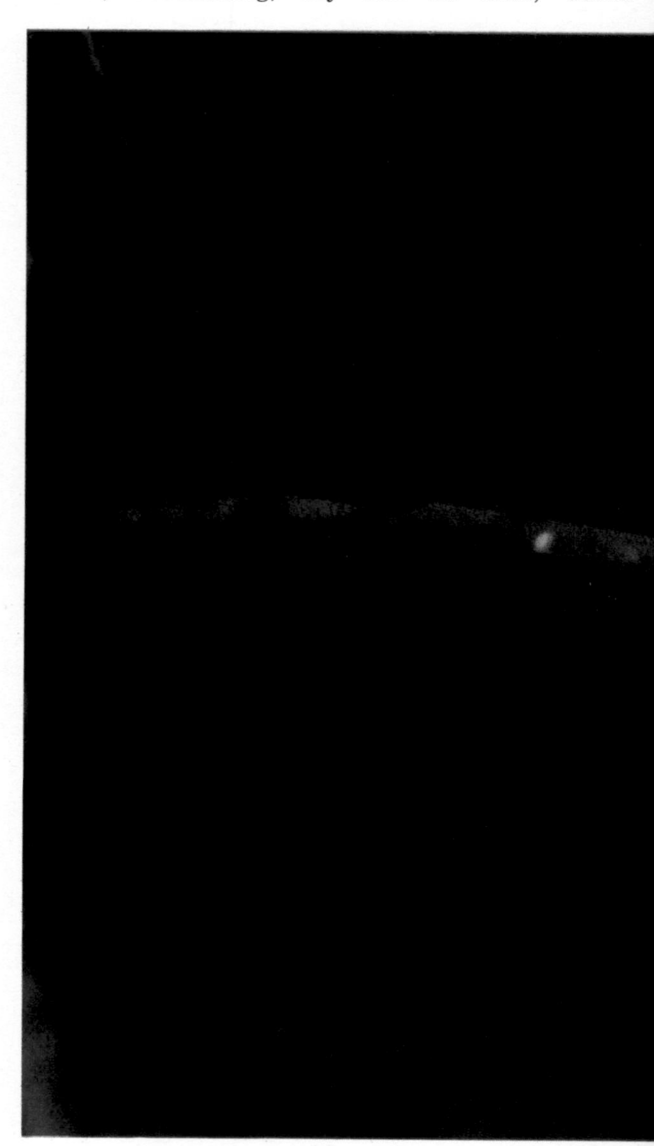

the land is still sodden, the weather is fine, but the dry wind soon dries out the land and withers the vegetation. The reversal of the monsoon winds acts like a seasonal sea breeze, but on a huge scale, so that the land is again parched as the long dry season progresses.

And so, unlike the equatorial rain forests, which are under the influence of the tropical low-pressure system all year round, the land north and south of the equatorial belt can support only such vegetation as can cope with the monsoon regime of alternating wet and dry seasons. The animals that inhabit this zone of monsoon forests or wooded savannas have also become adapted to seasonal fluctuations of food supplies. Many of them manage to survive only through a combination of migration and "birth control." For example, many weaverbirds are stirred into the breeding state only by the growth of green grass, which heralds the approach of a plentiful supply of seeds on which to feed their families.

The monsoon areas, with their marked seasons of flood and drought, are the habitat of millions of people who live precariously at the whims of the tropical low-pressure systems. Unfortunately, for reasons not altogether understood, the monsoons are not at all dependable. Failure

The red-eyed tree frog is a denizen of tropical rain forests, where daily downpours keep its skin moist. Some tree frogs even breed in rainwater pools trapped by cup-shaped leaves.

of the rains to appear at the usual time brings the prospect of famine and death to animals and people alike. Precipitation may be too low to make the land fertile for long enough to allow the crops to complete their growth.

In the early 1970s, a zone of despair stretched across 4000 miles of Africa, from Mauretania to Ethiopia, as a result of drought caused by successive years of failure of seasonal rains. This might have been part of a long-term climatic trend involving a change in habits of the equatorial low-pressure belt. At any rate, it meant starvation or near-starvation for millions of people in North Africa, and even in parts of India and Pakistan. On the other hand, the same period proved exceedingly fruitful for parts of Australia. With the tropical low-pressure area apparently displaced more to the south than usual, very heavy rains fell as far south as the central Australian deserts, which normally experience rain once in a human lifetime. Thus, slight changes in the movements and strength of the equatorial belt of air may make deserts bloom at the same time that forested and formerly productive regions become dustbowls.

Air drawn high into the troposphere by the equatorial sun divides, with one part traveling north and the other southward. As it cools, some of it slowly sinks, piling up in the region of latitudes 25° to 30° North and South, where it forms zones of high pressure. As the air sinks, it warms up slightly; and because it has very little water vapor—the vapor having been dis-

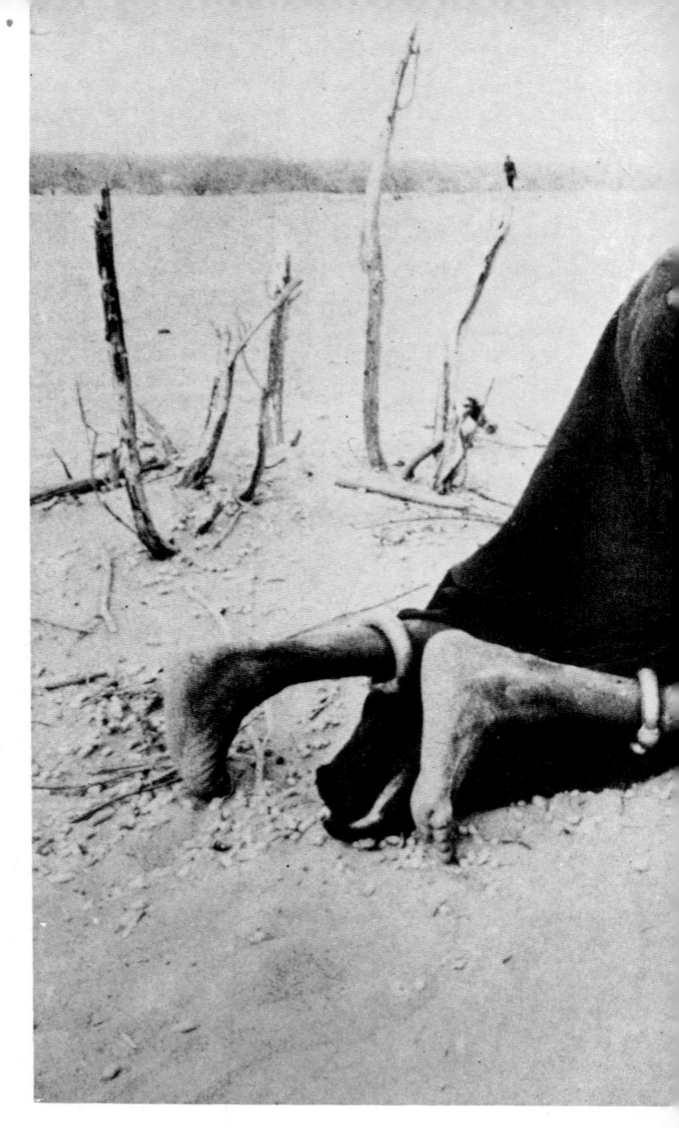

Above: an African mother grubs among drought-stricken shrubs for bran pellets dropped as cattle feed by a European relief organization. Here on the frontiers of Mali and Upper Volta just south of the Sahara Desert, shifts in the monsoon belt bring periodic drought, with fearful consequences to living things.

Refreshed by the rain, a bright green spinifex clump contrasts colorfully with bare orange sand—part of a desert scene in central Australia. Very slight temporary shifts of rain-bearing climatic belts briefly transform the drab colorless deserts into vivid, flowery meadows.

charged over the rain-forest and monsoon zones —the comparatively unsaturated atmosphere produces clear, dry skies. Sailors refer to these high-pressure fair-weather zones as the *horse latitudes* (the origin of the phrase is obscure).

Some of the air in the subtropic high-pressure areas of the horse latitudes flows back toward the equator, picking up moisture as it goes and forming the steady atmospheric currents known as the *trade winds,* which used to help along the boats that sailed between the continents. In the Southern Hemisphere, the trade winds blow up from the southeast, whereas those north of the equator come down from the northeast. Sometimes these trade winds, warmed and moistened by contact with subtropical seas, are the spawning grounds of cyclones, of which hurricanes and typhoons are very violent forms.

Such storms begin as low-pressure disturbances. When tepid air, heavily charged with vapor, is deflected upward into the atmosphere and into cooler latitudes, massive amounts of latent heat in the vapor are released by condensation, and this heats rising air, creating even more powerful updrafts. More and more humid air is sucked into the low-pressure center from the surrounding sea, and the further supplies of latent heat add fuel to the convection currents around the "eye" of the storm. As long as warm, moisture-laden air continues to be drawn into the center, the cyclone is self-sustaining, and it may grow into a hurricane, with winds of up to 200 miles an hour spiraling around the center and releasing water vapor as torrential rain.

When a cyclone is passing over the land, it can no longer suck up much wet air, and so its energy is dissipated, but very often not before

Local devastation may be caused by the low-pressure disturbances (such as hurricanes) that help to transfer heat from equatorial regions toward the poles. These scenes show the kinds of damage typically produced by intense low-pressure systems crossing the United States. Above: buildings set askew by a hurricane that struck Biloxi, Mississippi. Right: a family marooned by floods at Cape Girardeau, Missouri, where torrential rains overfilled the Mississippi River in the spring of 1973. Big storms moving up the Mississippi Valley flooded 11 million acres along a 1400-mile river frontage and rendered 35,000 people homeless.

cutting a swathe of destruction across the countryside. There are several spawning areas for cyclonic storms over the warm seas that are swept by the trade winds. About 10 hurricanes are brewed each year in the West Indies region alone; they sweep northward across Central and North America, generally in the summer and autumn, and their effects are often devastating. But perhaps the worst such storm on record was one that struck the Ganges Delta region of East Pakistan in November 1970. Flooding, tidal waves, and winds that flattened everything before them contributed to a death toll of more than 300,000, giving the storm the dubious distinction of being one of the worst natural catastrophes in recorded history.

And yet, surprising as it may seem, hurricanes and typhoons are not merely and entirely destructive, for they help to balance the world's heat budget. They contribute to the poleward transport of solar energy across the middle latitudes by sucking up vast quantities of warm air from the tropics and depositing the heat far to the north and south.

Other descending air masses over the horse latitudes are swept into the mid-latitude westerly winds. These are especially significant in the higher latitudes of the Southern Hemisphere (latitudes 40–50° s), where their course is relatively uninterrupted by land. The regions of the ocean where these tempestuous winds blow are known to sailors as the *roaring forties*. Both the trade winds and the westerlies assist birds in their migration, as we shall see later.

Where semipermanent high-pressure zones are situated over land, deserts tend to form. In Africa, for instance, the sinking air produces two areas of almost perpetually clear skies and low rainfall. The absence of clouds permits intense solar heating of the surface by day, and this both raises the temperature and evaporates any water in the atmosphere. Moreover, the air flowing out from such high-pressure areas blocks coastal rainstorms from penetrating inland, and the result is

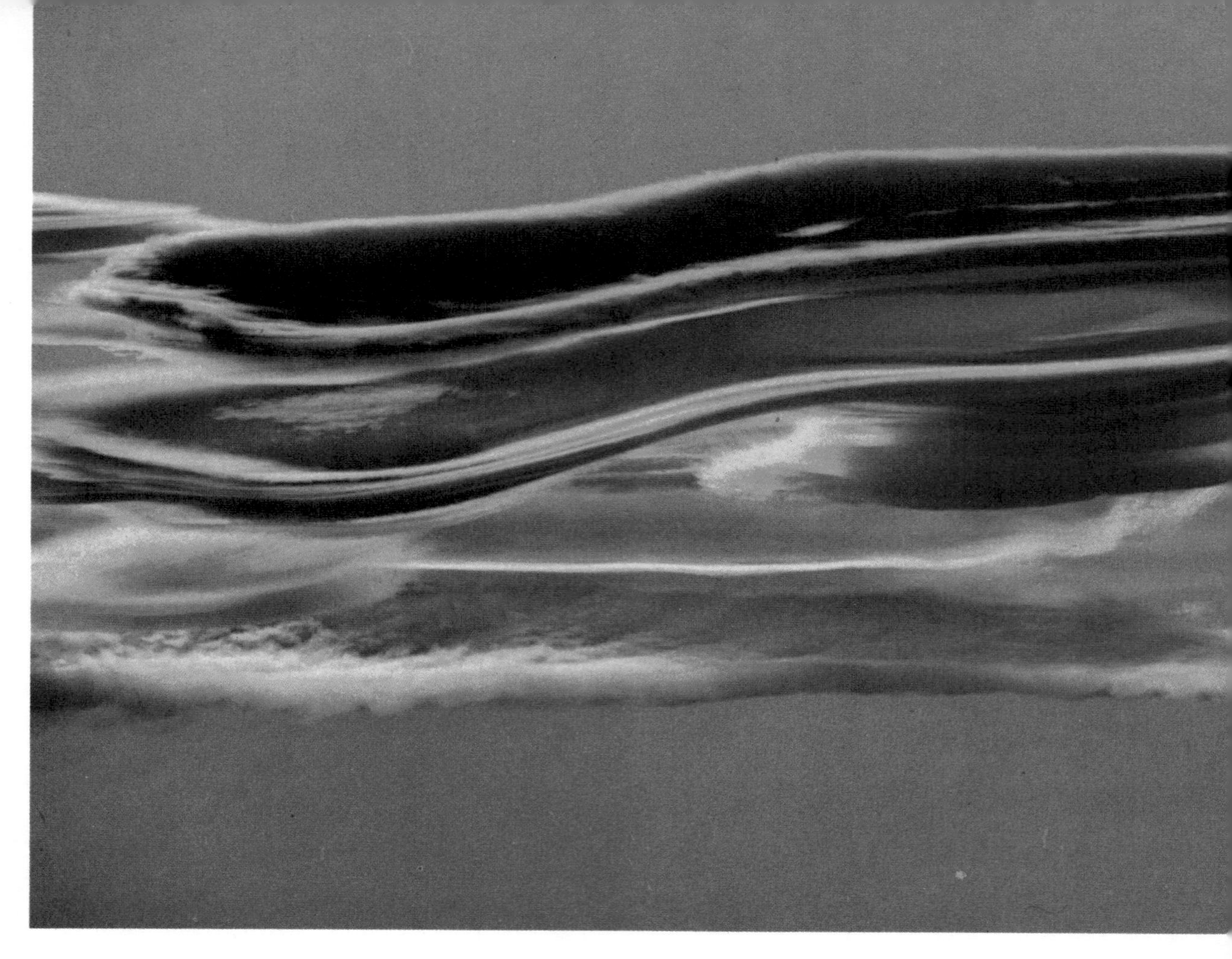

Four cloud formations typical of certain weather conditions. Above: lenticular or wave clouds, which form where air rides up in waves as it crosses mountains. Below: radiation fog, a cloud on the ground formed by moisture condensing in cool air sinking onto a cold land surface. Above right: altocumulus cloud (its base at 2400 to 6000 feet), a fleecy cloud formation that is usually a sign of fair weather. Right: a satellite camera's view of the seething vortex of clouds that reveals the violent formation of a hurricane.

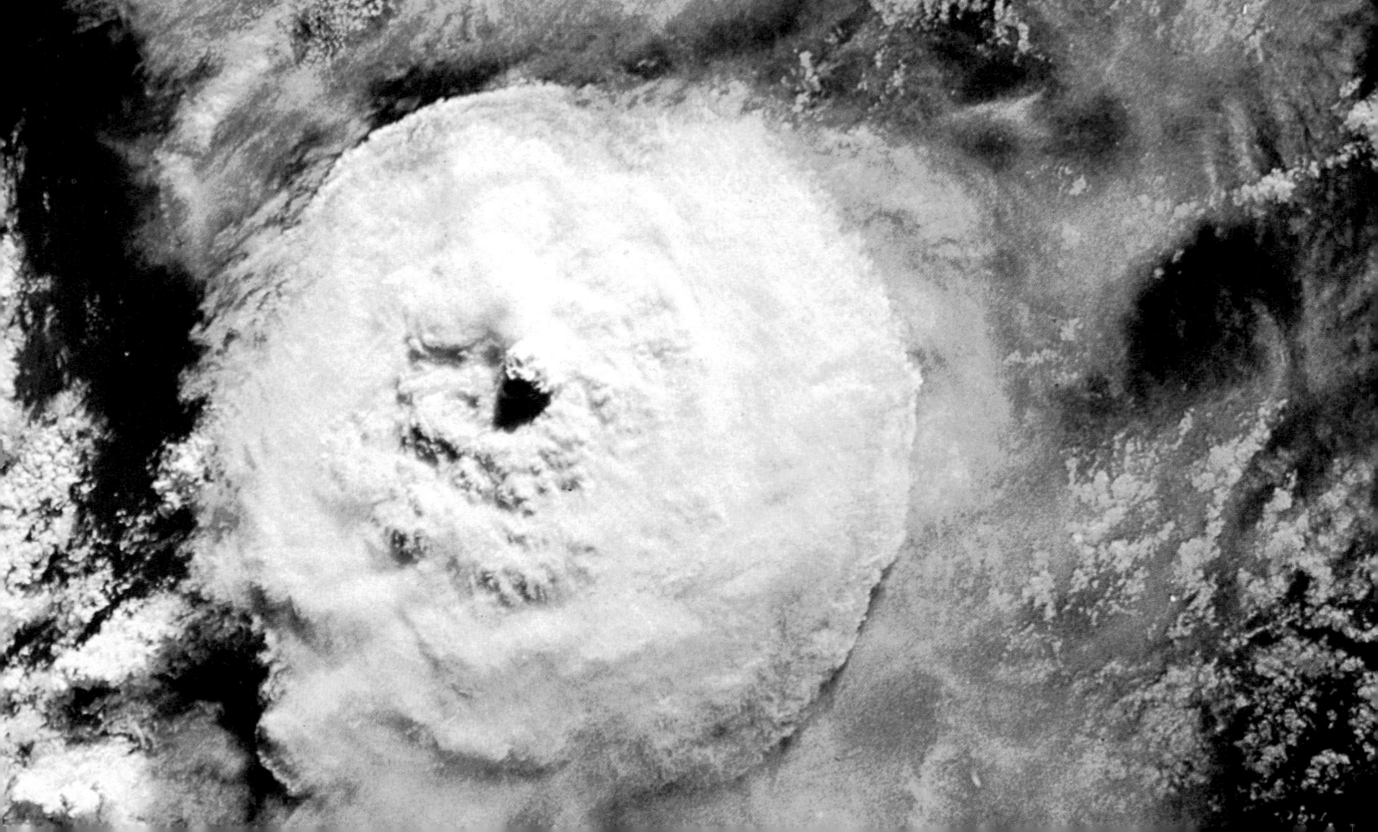

the aridity of the Sahara in the north and the Kalahari in the south. Other factors are involved in the creation of deserts, of course, but the subtropical high-pressure zones are a major contributing force.

The subtropical highs do not make survival easy for desert-dwelling animals and plants, whose lives are dominated by the need to keep cool and obtain enough water. This is one reason for the spines that often cover the surface of cacti: they cut down direct contact with the desiccating atmosphere. There is so little water vapor in the air over most deserts that 90 per cent of the daytime heat escapes into space at night, because there is not enough vapor to absorb infrared radiation. And so, unlike the stable stickiness of rain forests, the desert may be like an oven by day, with the temperature soaring tc 120°F, and like an icebox at night, at around 30°F.

This extreme cold, however, is vital to the survival of many desert animals and plants. When the air cools to below its dew point, it releases some of its moisture as drops of water. Although there is not much of it in the desert, the moisture can generally tide the plants and animals over the next hot day.

Obviously, the hottest part of the atmosphere is the boundary layer, where the air and earth meet. Most small desert animals tend to avoid the intense heat by burrowing, but it is an interesting fact that long legs are also a frequent desert adaptation. They are found, for example, in tenebrionid beetles and in such rodents as jerboas. Because the air temperature may be as much as 20 or 30 degrees cooler a few millimeters above the scorching boundary layer, it seems likely that long legs help to keep the little animals' bodies from burning up. In essence, though, it is not the heat of the atmosphere but its lack of moisture that poses the major survival problem for desert species.

Beyond the areas dominated by the subtropical high-pressure belts are the temperate zones. Temperate is an ill-chosen term, because both the Northern and the Southern Hemisphere temperate belts include a great number of widely varying climatic subdivisions. In many respects, the temperate zones are battlefronts, under the conflicting influences of warm, moist air from the subtropics and chilled, dry air from the polar ice caps. The circulation is simplest in the Southern Hemisphere, where only a very few major landmasses can complicate the situation with air of continental origin. The southern extreme of the world is dominated by a dome of ice 13,000 feet thick. Chilled, heavy air streams away from the dome in howling gales and eventually collides with the prevailing warm and moist westerlies that girdle the planet in the *roaring forties*. The

White Sands Desert, New Mexico, is in one of the world's two great desert belts, formed about 30 degrees north and south of the equator, where high-pressure regions predominate.

meeting between the two opposing air masses can bring foggy, squally weather to the area or can result in cyclones that go spinning off into the southern Atlantic, Pacific, and Indian oceans.

A more complicated state of affairs exists in the Northern Hemisphere, chiefly because the weather is much influenced by the vast continental areas that lie under the atmosphere. Then, too, the cold center of the Northern Hemisphere does not correspond to the North Pole. Tempera-

Overleaf: panoramic view of basic features in a depression moving left-to-right across a coast. Right: the depression's warm front (curved red line) approaches on a spring evening. Its warm air overrides cold air, raising the freezing level (the level above which atmospheric moisture freezes) and producing the ice-crystals of high cirrus cloud. Then come lower clouds including stratus and nimbostratus, respectively yielding drizzle and heavy rain. Left: next morning an approaching cold front (curved blue line) undercuts warm air. Moist warm air forced to rise in turbulent cumulonimbus cloud cools, producing an anvil cloud formation, and triggers a thunderstorm's rain-bearing downdrafts.

COLD FRONT

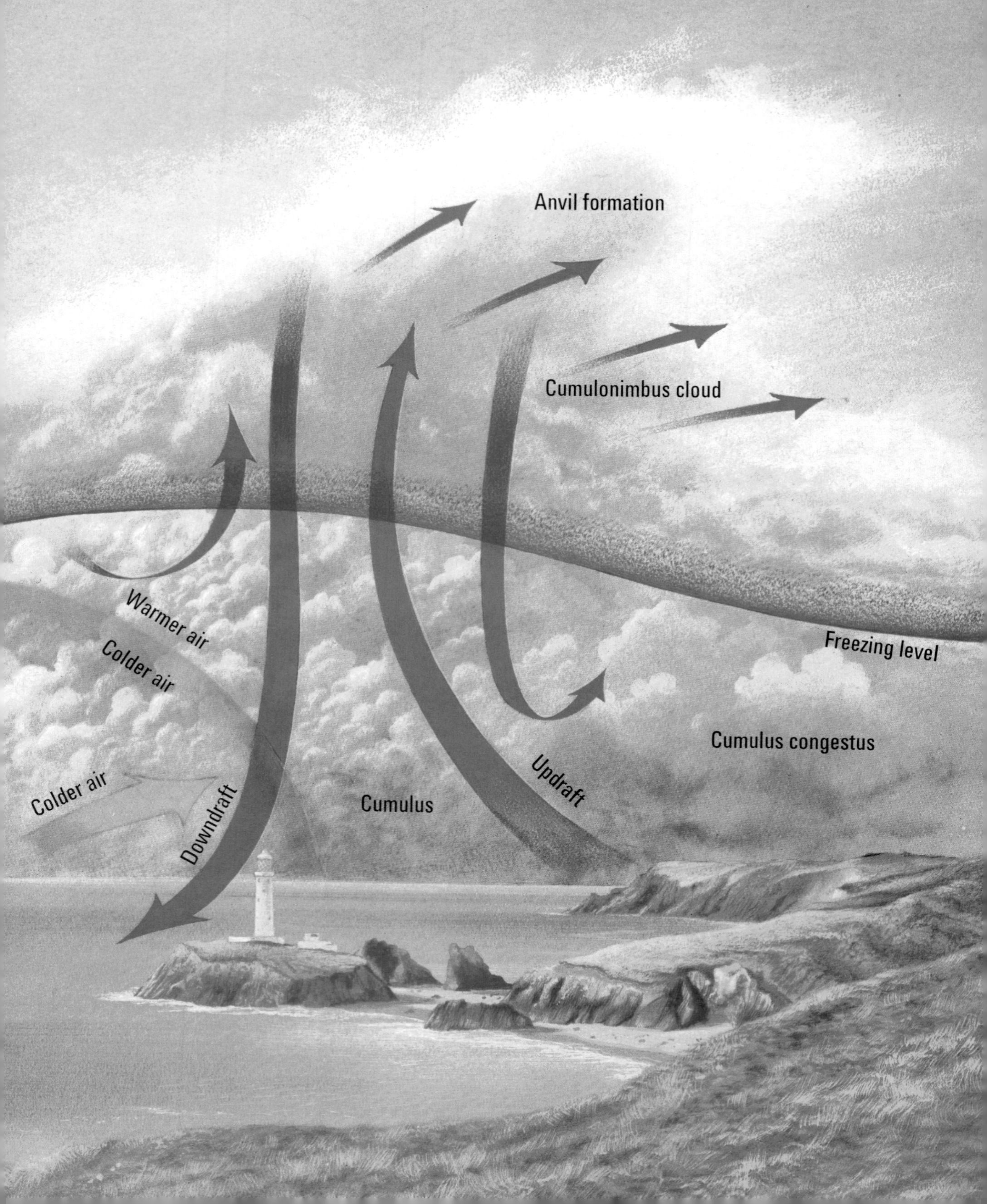
Anvil formation
Cumulonimbus cloud
Warmer air
Colder air
Freezing level
Cumulus congestus
Updraft
Colder air
Cumulus
Downdraft

WARM FRONT

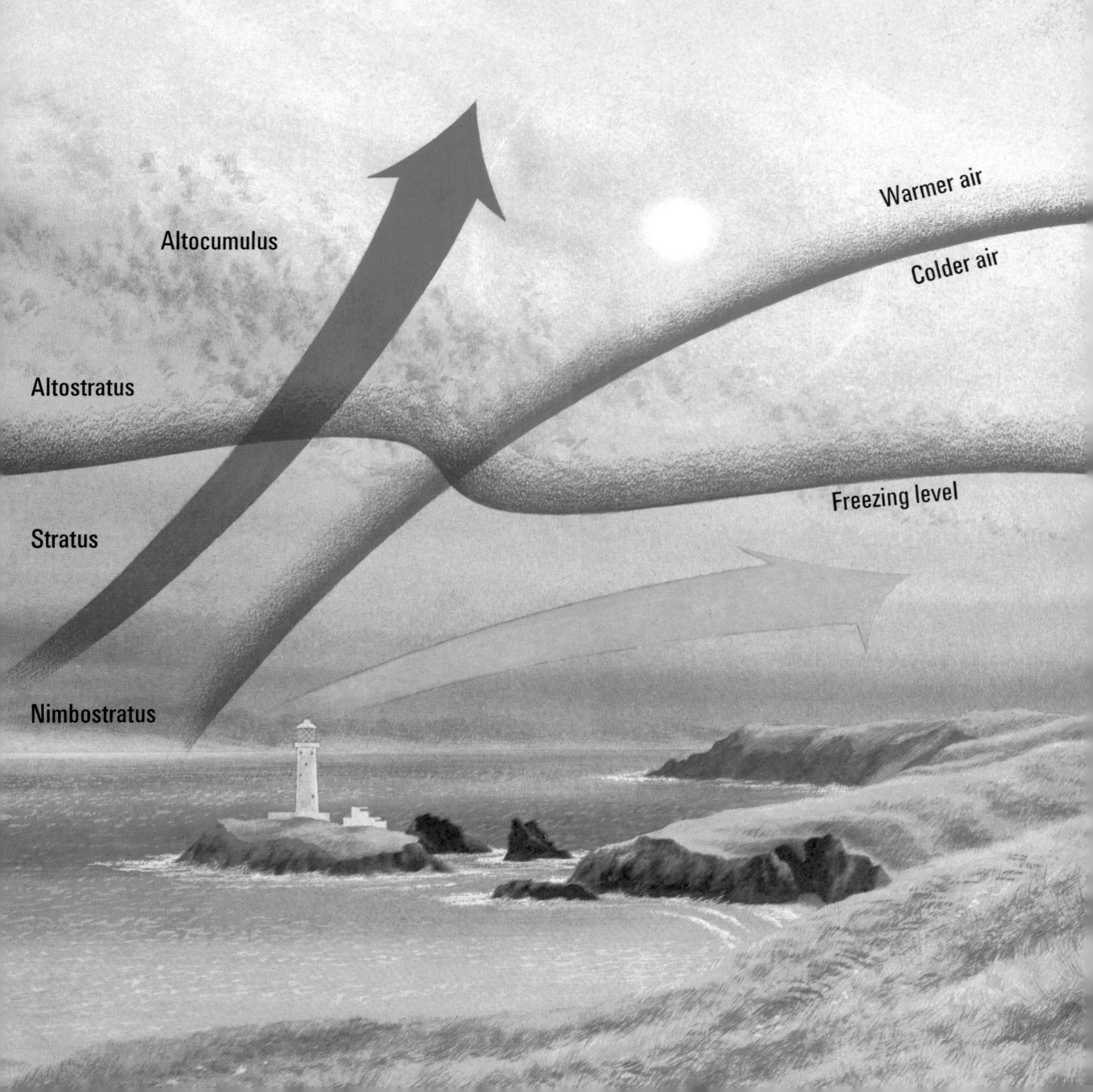
Cirrus
Cirrostratus
Warmer air
Altocumulus
Colder air
Altostratus
Freezing level
Stratus
Nimbostratus

Left: tornado near Freeman, South Dakota. Tornadoes are violent whirling winds that begin with a funnel of air spinning down from a dark cumulonimbus cloud. On reaching the ground it stirs up a great cloud of dust. As it travels it cuts a path perhaps 300 yards wide by several miles long. Houses explode and vehicles, people, and animals whirl hundreds of feet through the air. More than 200 tornadoes annually hit the United States, mainly affecting Middle Western States in the spring months.

Right: fleecy seed-bearing heads of cotton grass dot an Alaskan bog. Only low-growing plants able to survive bitter winds, long, dark winters, and permanently frozen subsoil thrive in the treeless wastes of such tundra regions. The world's far north and south receive relatively little of the airborne heat that flows northward and southward from the permanently hot equatorial zone.

tures at the South Pole reach as low as $-120°$F; but at the North Pole the thermometer rarely records a temperature under $-28°$F, because the pole is situated on sea ice only a few feet thick, and the air above it gets a lot of heat from the seawater. Actually, the earth's northern cold center is situated just within the Arctic Circle at Verkhoyansk in Siberia, 1500 miles from the North Pole, where a low temperature of $-100°$F has been recorded. And other centers beneath cold descending air occur over Greenland, Iceland, and Northern Canada.

The warm and cold fronts so familiar to those who watch weather forecasts represent the boundaries between air masses of different origin, and they herald short-term changes in weather. A warm front occurs when warm, moist air advances, rising above the cold air and displacing the cold-air mass. As this happens, clear skies, typical of high pressure and unsaturated air, slowly give way to clouds and rain, for condensation takes place as the warm air rises and cools.

When a cold front approaches, on the other hand, it usually means that the leading edge of a dense mass of cold air is knifing its way underneath warm, moist air and pushing the warm air upward. This normally results in the kind of massive clouds that we associate with thunderstorms. But a particularly violent frontal situation can spawn a tornado—perhaps the most terrifying and destructive of all winds. Such a disaster can occur when a cold front moves so quickly that, instead of forming a wedge beneath the warmer air mass, it sweeps over the top, setting up a very unstable situation. The warm, lighter air, temporarily trapped, forces its way up through the cold layer in ominous, shaggy funnels, which are given a twist by air rushing into the violent updraft. Rotating upward at 200 to 300 miles an hour, the funnel cloud of a tornado usually moves across the countryside at a speed of up to 40 miles an hour, and its immense power can lift cars, crush buildings, and uproot trees along a 10-mile pathway. Such whirlwinds can happen almost anywhere, but most of the violent ones occur in midwestern America, when warm,

moist air moving up from the Gulf of Mexico is overlain by cool, dense air from the west or northwest.

The interaction between frontal systems broadly determines the nature of temperate-zone vegetation. As a general result of this interaction, temperate-zone regions can expect moderate rainfall throughout the year. And so, if it were not for the handiwork of generations of farmers, much of the landscape on the edges of the North American and Eurasian landmasses would be clothed in forests, with a mixture of deciduous trees and evergreens. By the time air reaches the center of the continents, however, much of it has usually lost its moisture; as a result, the vegetation is less rich, and some type of natural grassland is likely to prevail.

Beyond the temperate zones, as the temperature gradually drops, conditions tend to favor the survival of spruce and pine, and mixed forests slowly give way to the entirely coniferous taiga. In the Northern Hemisphere, the taiga merges into the arctic tundra, where icy winds allow only dwarf willows to grow, together with plants that can keep their heads out of the chilling, drying winds by sprawling close to the sun-warmed surface. Despite the frigid winds and the long, dark winters, however, about 400 species of plants make up the tundra carpet; and during the short but continuously light summer, these provide a living for numerous insects. As autumn approaches and brings temperatures as low as −20° F, some of the tundra insects secrete glycerol as a kind of antifreeze to protect their bodies. Many rodents and birds, and even animals as large as the musk-ox and the caribou also survive the cold blasts of the tundra, but the cold-blooded reptiles and amphibians are absent.

Although the earth's life zones are primarily dependent upon varying atmospheric conditions that prevail at different latitudes, the picture is obviously made more complicated by geographical features such as mountains and ocean currents. For instance, when a mountain projects thousands of feet into the rarefied regions of the atmosphere. the cold, dry air and fierce winds

A polar bear in a snowy Spitsbergen setting. Its warm yellow-white fur (camouflage against the snow), swimming ability, and savage jaws and claws equip this large predator to hunt and to kill the seals with which it shares the climatic rigors of the Arctic region.

California's giant redwoods obtain summer moisture as mist deposited by damp sea air cooled by a Pacific ocean current.

will obviously have a greater effect on plant and animal life than will the mountain's distance from the equator. This is not to say that latitude loses its importance where mountains are concerned. Latitude and altitude together mold the nature of the environment. For example, trees cannot grow satisfactorily where the average monthly temperature is below 50°F; on the equator the line beyond which the atmosphere can no longer support trees may be drawn at 13,000 feet, whereas the timberline in Scandinavia may be below 2000 feet.

For examples of how ocean currents have a modifying influence on the climate we need only look at the California coast and the British Isles. Cool Pacific water flowing southward past California in the spring and summer months causes the warm westerlies to precipitate their moisture as cooling mist. This supplies the giant redwood forests in the coastal regions with virtually all their water; the moisture collects on the leaves, then drops to the ground, where the roots soak it up. In the northern Atlantic, the effect of the warm Gulf Stream on the British Isles is similarly remarkable. London's climate, for instance, is much less extreme than the frigid winters and sultry summers of New York, and yet London is situated 10° of latitude farther north, on much the same parallel as northern Newfoundland.

Cities in general can cause quite marked alterations in local climatic conditions. As compared with the nearby countryside, any large city is likely to be not only dirtier but cloudier, warmer, and wetter. The concrete and asphalt surfaces of a city absorb and hold the heat of the sun. As night approaches and some of the warm air rises, more humid air flows in from surrounding areas. Dust in the atmosphere—around 65 million particles per cubic yard, as opposed to some 100 salty particles per cubic yard over the oceans—provides vast numbers of nuclei for condensing water droplets, possibly making it easier for clouds to form like a blanket over the city streets.

Cities tend to be warmer than the countryside by an amount that varies from 5° to 25°. Spring flowers open earlier in towns, and some birds that normally feed in the countryside choose to fly into warm city centers to roost. Starlings are perhaps the best-known city sleepers, but urban roosting in wintertime has been recorded for several other kinds of rural feeders.

So far, then, we have seen something of how the atmosphere, which is always in a state of flux,

creates the climate, and how weather patterns are formed by complex movements of the air in conjunction with other geographical factors. It may seem at this point as if all life is at the mercy of the winds. But that is only a half truth. Many living creatures make good use of the moving air to keep themselves and their species alive. The wind often serves as a powerful aid to dispersal of both animals and plants. There are a number of ways in which plants harness the wind as a means of spreading their seeds. Many members of the amaranth family have an unusual method. When its seeds are ripe, the plant dries up into the shape of a ball, breaks away from its roots, and rolls before the wind, dropping small fruits as it goes. The best-known of such "tumbleweeds" is *Amaranthus graecizans*, which is indigenous to the western United States but has spread quite far to the east.

This method of dispersal is practicable only for

Left: struck by a water droplet, the ripe fruiting body of a puffball releases a fine cloud containing millions of minute spores. Thus propelled into the atmosphere, such spores float aloft on the lightest breeze. Some drift many miles. But only one or two land in a suitable place and grow into new puffballs.

Right: wind-blown tumbleweed on the move. As the ball-shaped plant rolls and bounds across desert or prairie, ripe fruits drop off and their seeds tend to root where they fall. Moving air thus helps to spread these plants, whose seeds would otherwise all drop together, producing seedlings that might overcrowd each other.

species that inhabit open deserts or prairies. Plants that live packed together, or hemmed in by taller kinds, may also employ the air as a transport agency, but in a different manner: they produce tiny spores, pollen grains, or seeds that can be carried away by the slightest breeze. Even apparently clean air may contain enormous numbers of such microscopic structures. The fungi in particular liberate spores into the air in prodigious numbers. A giant puffball can reputedly produce 7,000,000 million spores, and a large bracket fungus can liberate 30,000 million spores into the air every day over a period of six months.

Many fungi project their spores explosively into the air, to give them a better chance of reaching air currents beyond the stagnant boundary layer. It is little wonder that fungal spores are found almost everywhere in the atmosphere. Spores have been trapped in a stratospheric balloon at an altitude of over 36,000 feet. But closer to the ground, the concentration is higher, of course. In a recent experiment an airplane, flying at a speed of 85 miles an hour 1000 feet above some Canadian wheatfields infected with a type of rust fungus, carried a greased slide that was exposed in the slipstream. In 10 minutes the slide picked up 24,200 spores. Only 7560 were intercepted at 5000 feet, 108 at 10,000 feet, and a sparse 10 at 14,000 feet.

The dustlike spores are so small and light that those carried to great heights by turbulence or rising currents of warm air may be transported hundreds, or even thousands, of miles by the winds before they sink to the ground. Wind direction plays a key role in determining where the spore showers will occur. Some years ago, when black rust was rampant in American wheat-growing regions to the south of latitude 40°N, strong southerly winds blew up during the first week of June. By June 10 the black rust had infected all susceptible wheatfields up to 600 miles north of the 40°N parallel.

Sufferers from hay fever are only too familiar with the fact that at certain times of the year the air contains great quantities of pollen. As it is wafted along, it no doubt makes some people sneeze, but it also helps its parent species to survive and provides direct nourishment for some other forms of life. The ice flea, which lives on the surface of snow and ice, feeds almost exclusively on wind-blown pine pollen. Pollen from lowland areas makes a significant contribution to the diet of insects that live high up on mountain ranges, to which the wind has carried it.

The minuscule dustlike seeds of orchids are easily dispersed by moving air. But plants with heavier seeds have evolved various seed shapes that are superbly designed to catch the wind—among them, seeds that look and act like vanes, screw-propeller blades, disks, or (as in pines and sycamores) wings. These last, the so-called flying fruits, are liberated into the air from high up on their parent tree; and through the action of their one, two, or even three wings, they rotate and

Three kinds of seed dispersed in air. Left: "winged" sycamore seeds, which rotate and drift to earth. Below: an enlarged section through the ripe flowerhead of a Tweedia *(related to the milkweed family). Silky floss parachutes bear its seeds away on the wind. Right: an alfalfa plant, showing the coiled pods that will later split and liberate the ripe seeds into the atmosphere.*

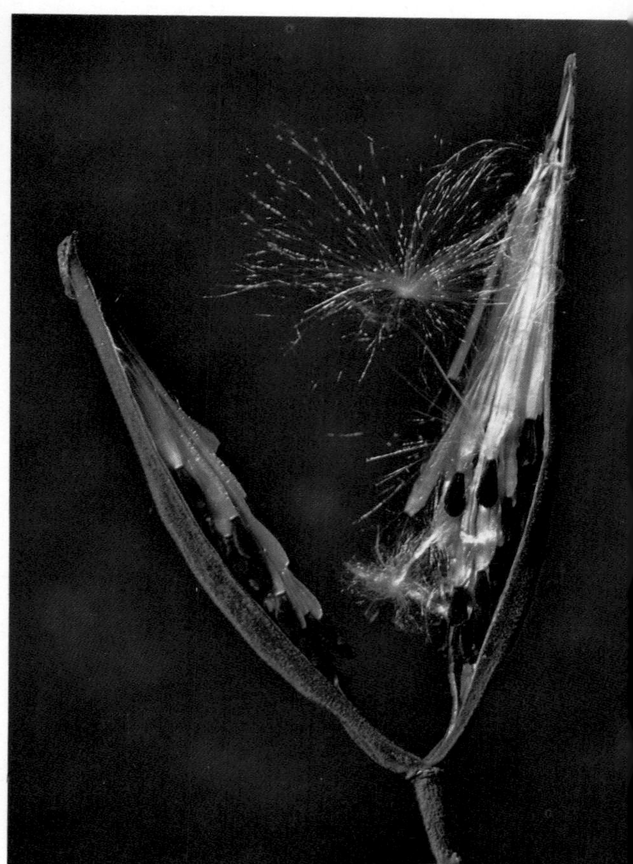

drift down to the ground like tiny helicopters.

Species such as the dandelion have developed seeds equipped with little umbrellas or parachutes, which help the wind to blow them as far as possible from the parent plant. Such parachute mechanisms are often sensitive to humidity. The hairlike structures that form the "umbrella" cling together heavily in damp air, but separate and open out for sailing when it is dry. This makes the dispersal structure most effective when weather conditions are favorable to wind carriage.

In principle, the dispersal methods of the plants are not unlike the technique employed by migrating money spiders. These tiny arachnids sometimes live in incredible numbers—perhaps 25,000 to an acre—near the ground. In the autumn, they climb upward to some pinnacle such as a flower or wall, then face the wind, secreting silk as they do so. The air currents catch the silk and draw it out into fine strands, until the drag is sufficiently powerful to lift the spider and waft it away. This is a very effective means of transport. Money spiders traveling before the wind on their gossamer drag lines have been picked up on ships' masts over 100 miles from land, and aircraft have intercepted them at an altitude of 14,000 feet over the American mainland.

The money spiders' ability to hitch a ride on the winds gives them a good start in the race to colonize new lands. There is a small arctic volcanic island called Jan Mayen Island, separated from Greenland and Iceland by about 350 miles of cold, windblown sea. Yet four kinds of money spider have managed to ride the wind across the Arctic Ocean and settle in this isolated place.

Wind power is perhaps most gracefully exploited—among the invertebrates, at least—by the Portuguese man-of-war. Like the galleons of old, these formidable creatures slice through the water by means of an oblique sail-like projection on top of a submerged substructure. Although it looks rather like a jellyfish, the Portuguese man-of-war is in fact a colony of small organisms. All that can be seen on the surface is the beautiful blue-and-mauve sail, composed of a bladder topped by a crest; in the water, long tentacles trail behind, and they carry the stinging cells with which the creature attacks and captures its prey.

The Portuguese man-of-war has no means of independent propulsion, but relies completely on water currents and the wind. It cannot alter the set of its sails to tack, and so the line that one of these conglomerate animals takes with a following wind depends upon whether it is equipped with a right- or left-handed sail. As it happens, each brood of these siphonophores (to give such colonies their technical name) contains approximately equal numbers of right- and left-sailed individuals, and the prevailing winds tend to separate them out into fleets of one or the other type. This obviously has survival value, because, if faced with a potential stranding situation, half the endangered colonies would be blown safely away from the beach.

Nowhere, of course, is the impact of wind and weather on traveling better illustrated than in those superb aeronauts, the birds. When they are not nesting, many seabirds travel on the steady, reliable trade winds and westerlies that blow across so many regions of the oceans. For example, a whole range of albatrosses and shearwaters circumnavigate the globe before the winds of the roaring forties. If they do not need to battle against the force of air, birds can conserve a great deal of energy, and those that have evolved as sophisticated gliders make extraordinary use of wind power. As the breeding season approaches, the albatrosses and shearwaters simply peel off from their westerly wanderings to join their rookeries, which are situated on numerous islands within easy reach of the roaring forties.

The pattern of migration taken by the short-tailed shearwater, or muttonbird, is dictated by both the availability of food and the oceanic wind system. After nesting chiefly in the region of Tasmania and the southeast corner of Australia, the population travels on a looped course that takes it over the equator, past China toward the

Aleutian Islands, then by way of the California coast westward across the equator, passing north of New Zealand, back to Australia. In chasing the sun northward, the birds make full use of the southeast trade winds, which help them up toward the China Sea; there they pick up the North Pacific westerly air stream, which they can ride until they reach the North American coast. Once through the equatorial doldrums, the shearwaters are assisted on their last leg across the Pacific by the southeast trade winds again.

Like the shearwaters, and like human travelers as well, many other birds have found (through evolution) that the direct short course is not necessarily the quickest and most economical way to travel. There is evidence, from tracking migrating birds by radar, that there are places where even land birds seem purposefully to head

Winds deeply influence much insect and bird migration. Left: monarch butterflies wintering in California; winds may aid their northward flight in the spring. Below left: a light-mantled sooty albatross nesting on a subantarctic island—a pause in its lifelong windblown circling of the Southern Hemisphere.

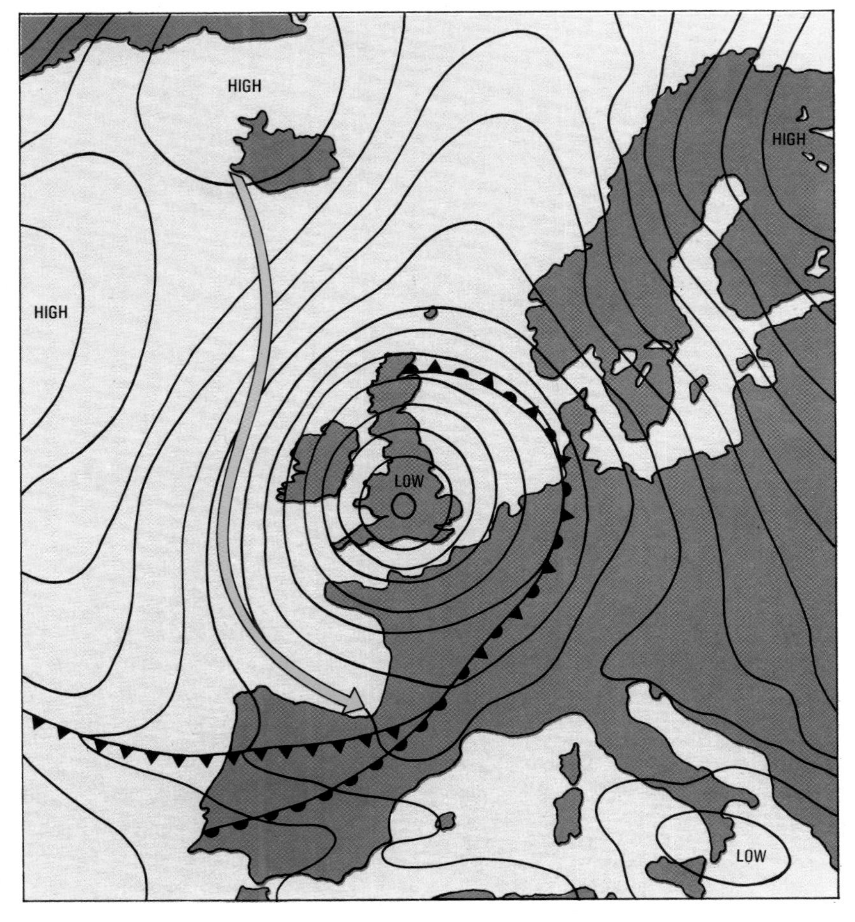

Right: migratory route of redwings. They migrate south from Iceland to Spain by riding on the winds blowing counterclockwise around a low-pressure system centered on the British Isles. Because atmospheric systems can swiftly move or dissipate completely, winds may shift or drop, and many birds are lost at sea.

out over the oceans, in order to get assistance from trade winds. For example, all the birds that migrate southward from North America to South America could in theory reach their destinations either by island-hopping across the Caribbean or by keeping entirely over land, across Central America. Yet some of them leave the coast of the United States and fly southeastward into the North Atlantic, across the quiet horse latitudes, until they run into the northeast trade winds coming down from south of the Azores; and this easterly airstream wafts the birds into Venezuela or the Guianas. Despite the possible danger of coming down in the sea and drowning, natural selection has obviously favored birds that take the risk. The reason, no doubt, is conservation of energy coupled with the fact that the sea route is probably quicker than an overland passage battling against head winds would be.

The need to get the maximum help from the wind may be one reason why many migrating birds fly so high: wind speeds generally increase with altitude. Few people realize just how high some flocks soar, and quite regularly, too. Chaffinches leaving the coast of Holland with a northeasterly tail wind climb to about 5000 feet, and red-backed sandpipers arriving on the east coast of England after a night journey from Scandinavia may be around 6000 feet. Peak altitudes of 21,000 feet have been recorded for some migrating flocks of birds. And shore birds migrating directly across the Atlantic from North America to eastern South America occasionally seek the windy heights of 20,000 feet.

Birds need to see the sky in order to navigate. With the sun or star beacons obscured by mist or cloud, they become disoriented, and if land birds are caught in bad weather out of sight of land, they simply drift downwind. Many species therefore start their autumnal migratory flights only when there are clear skies and light, favorable winds. Such ideal conditions are often only local, however. Because of the turbulent nature of the prevailing weather in the North Atlantic, any land-bird migrant, particularly from northern latitudes, has to be prepared for a very long flight

Even such strong migratory fliers as these blue geese and snow geese (right) delay spring migration until wind and visibility favor the long trek northward to their Arctic breeding grounds. Blue geese then leave their winter bases around the Gulf of Mexico. Snow geese similarly migrate northward in North America and Eurasia. Sometimes birds start their journey in favorable weather that suddenly turns foul. Storms forced these starlings (below) to rest on a passing ship far out at sea.

over the sea, at the mercy of cyclonic winds. Accordingly, long-distance travelers need stamina and extra fuel.

That, no doubt, is why such far-north birds as the redshanks, redwings, wheatears, and merlins of Iceland and Greenland are larger than members of the same species based several hundred miles farther south. They also have proportionately larger wings, capable of providing the extra thrust and lift needed to carry their additional weight of muscle and fat. Although these birds may have to fly "blind" in the teeth of an Atlantic gale, they rely for survival upon their ability to ride fast and far on favorable westerly winds.

But although the westerlies may be of assistance to migrants from high mid-Atlantic latitudes in autumn, they can be only a hindrance during the return spring journey. Luckily, the storm track weakens during spring and early summer, and so northward-traveling birds are allowed a respite from the steady succession of cyclones. Many geese that spend the winter in Europe and North America, such as the white-fronts, breed in Iceland and Greenland. While frontal cyclonic weather persists, with rain, cloudy skies, and westerly winds, the geese remain on their winter feeding grounds. With the onset of calm conditions, they make their migratory flight northward, possibly with a southerly tail wind all the way to Iceland. The prevailing westerlies over the Atlantic make it extremely difficult for birds of European origin to cross over to America, whereas American migrants sometimes get a wind-assisted passage to Europe. When an American pectoral sandpiper or olive-backed thrush happens to be blown off course into the Atlantic while attempting to migrate southward across the United States, it can easily become caught up in the storm belt, which may then carry it quickly across 2000 miles of sea.

This sort of transatlantic crossing rarely takes place in the springtime, despite the fact that large numbers of northbound American migrants are swept out to sea off Florida by sudden outbreaks of tropical storms in the spring. The North Atlantic storm track is generally too weak at that season to carry the disoriented birds all the way to Europe.

A few European species have managed the east-west journey, but only rarely, when the winds have been especially favorable. The westward flight is comparatively easy in the Southern Hemisphere, however, where the trade winds assist their passage. One species of egret—the so-called cattle egret—evidently crossed from Africa to South America with the help of the trade winds. The first cattle egrets in the New World were sighted in South America at the beginning of the 20th century, and since about 1930 the species has spread explosively.

Obviously, then, all flying creatures—not just birds, but man-made machines and such insects as butterflies and locusts—depend on the weather as well as on their wings to carry them along. But the variations of wind and rain merely *influence* flight; they do not create it. The conquest of the air by animals involves much more than the ability to "float" on moving atmosphere. It involves more, too, than the mere acquisition of a pair of wings. As we shall see in the next chapter, the act of flying requires a very great degree of engineering skill.

Westerly winds prevent European birds from reaching the Americas. The cattle egrets (below) invaded America on favorable trade winds from Africa. They were first sighted in South America at the beginning of the 20th century. But even contrary winds are unable to stop the 11,000-mile migration of the arctic tern (left), which commutes between the Arctic and Antarctic regions regularly.

Conquest of the Air

Squirrel monkeys may hurtle 20 feet from one branch to another as they roam through the South American forests in search of food. Monkeys simply jump. But some mammals, most birds, and many insects truly fly or glide through the great sea of air.

Mastery of flight must rank as one of the greatest achievements of evolution, for even at the bottom of the atmospheric ocean the air is so thin that it offers little support against the tug of gravity. A fish can float easily because the mass of water it displaces provides an upward thrust that more or less balances the downward weight of its body. A terrestrial animal gets no such help from buoyancy; once the land-dweller loses contact with the ground, the problems of resisting the gravitational pull of the earth become enormous. Not only does flying need good body design, skill, and a great deal of power, but it exacts exceptionally severe penalties for clumsiness or technical failure. After all, if a running man or mouse stumbles, the physical consequence may be no more serious than a graze, but a cramped muscle or uncontrollable stall of an animal in flight will almost inevitably result in a body-shattering crash landing.

Defying gravity is not, therefore, something to be undertaken casually. In becoming airborne, the flying animals took living materials and design to their evolutionary limits. The problems solved are those that plague every designer of aircraft. All the problems center around one big question: how to keep the weight of the airframe to an absolute minimum without at the same time sacrificing strength and power? For "excess baggage" is costly in fuel and resources if it has to be lifted and dragged through the air, whether in the hold of a jumbo jet or in the body of a sparrow.

Despite the difficulties, so many animals have become aeronauts that we can justly conclude that the rewards of having the freedom of the skies are very great. Some are obvious. Being literally one up on earthbound enemies is a powerful selective advantage of flight. A house sparrow with its wits about it is hard for even the best of mouse-catching cats to capture, but a flying predator can often make an easy living by plucking its food from the ground, or even out of the air. In flight, too, new resources can be tapped that might otherwise be inaccessible. For example, the nectar and fruit exploited by flying animals may well be out of reach of others.

Such formidable barriers to earthbound travelers as oceans, mountains, and deserts are easily crossed by winged animals, and so flight increases the potential for long-distance travel. Moreover, the speed of flight permits the flying creature to cover tremendous distances in short periods of time. It is inconceivable that a walking or running animal could commute annually between, say, Denmark and South Africa by way of the Middle East; yet white storks do so.

Less immediately obvious are some of the advantages that the flying animal gains through its ability to take the shortest route between two points. Consider what can happen if a creature must find its food in the canopy of trees. A good leaper, such as a squirrel or a sifaka, or a swinger, such as a gibbon, can probably manage quite well most of the time by springing from one tree to another. But even these beasts must sometimes descend to the ground in order to reach a new tree, and climbing up and down involves the expenditure of much energy. It also involves exposure to the danger of ambush. So it is not only easier to glide or fly from tree to tree, but also more economical of muscle power, and safer.

With so many benefits, it is little wonder that a great many creatures have taken to the air. In addition to birds and insects, a number of fish, amphibians, reptiles, and mammals have independently evolved aerial designs. At least one mollusk (a "flying" squid) has done so, too. Not all of them have been entirely successful in conquering the air but all have had some success.

Aerial animals belong to one of two categories: sky divers, and the true aeronauts, which fly under power. Any very small animal can be a sky diver and can perform death-defying leaps from great heights. The reason is simple. Under the pull of gravity, an object will continue to accelerate downward until the drag of the air flowing past it acts as a brake and keeps it from falling faster. When the sinking speed becomes constant, the falling body is said to have reached its terminal velocity. A little creature such as a wood louse or a mouse never falls fast enough to hurt itself badly when it hits the ground, no matter how far it has dropped, because it has a very large drag-inducing surface area compared with its weight. That is why Brazilian tree frogs

that were dropped from a 140-foot-high tower during the course of a scientific experiment all landed unscathed.

Unfortunately, free falling becomes progressively more hazardous for animals as their size increases, because greater weight is not matched by an equivalent increase in body surface. A man plummets earthward at a terminal velocity approaching 100 miles an hour unless he can equip himself with many square yards of extra surface without adding appreciably to his weight. This is the principle behind the design of parachutes. The large surface area of a relatively lightweight billowing parachute acts as an air brake, raising the drag force enough to allow a slow descent and, as a result, a soft landing.

The bushy tails of such arboreal mammals as squirrels, tree shrews, possums, and some lemurs —all of which are large enough to suffer physical damage in a free fall from a high branch—are used as built-in parachutes. When fully fluffed the tail retards its possessor's fall, and it also helps squirrels to make broad leaps between branches without losing too much height in the process. These furry air brakes have to be particularly effective in the larger and heavier species; some of the giant squirrels of the Far East have two-foot-long tails, which are half as long again as their bodies. Nevertheless, all good things must come to an end. There is a point on the size scale where an efficient personal parachute would be so cumbersome as to become a liability rather than an asset. A parachute tail big enough to break the fall of an orang-utan

Ability to fly enables flamingos to migrate often between Africa and Europe. Flight also allowed flamingos disturbed from the Caribbean islands to make their home on mainland Mexico.

would need to be the size of a small tree.

Amphibians have evolved their own parachute brigade in a group of chiefly nocturnal tree frogs that live in the rain forests of Malaysia and Borneo. When Westerners first heard about frogs that were capable of flying from the treetops, such stories were credited more to the potency of the oriental imagination than to zoological accuracy. But the stories turned out to be true. A number of tree frogs are equipped with extra-long webbed toes. When airborne, the frog splays its toes and unfurls four tiny parachutes, one on each foot. Furthermore, it "draws up" its body to produce a concave belly, which may increase the lift. These tree frogs can glide as much as 50 feet from an upper branch of one tree to the base of another. Like other tree frogs, they gain height by climbing, which they do with the help of suckers on their toes.

If gliding frogs seem strange, the flying snakes of Southeast Asia, India, and Sri Lanka are even stranger. A snake seems far removed in design from the animals that have become adapted to flight; and yet the flying—or, more appropriately, gliding—snakes are quite accomplished at moving across gaps in the canopy. The golden tree snake is a venomous species that grows to a length of about three feet and that hunts by day in the treetops. When gliding, the snake launches itself into space, expanding its ribs to flatten the body, thus increasing its surface area. At the same time it throws its body into a series of moving lateral waves, so that, when seen from below, the snake resembles a big eel swimming through

Long, muscular arms help the white-handed gibbon to swing easily through the trees. Yet the gibbon's inability to fly exposes it to dangers that airborne animals are able to avoid.

the air. The tail region is not flattened but remains cylindrical; to judge from the way it is waved around, it may well act as a stabilizer.

Surprisingly enough, this snake glides down headfirst at a comparatively shallow angle. By contrast, if a nonflying snake of similar build is ejected from the top of a tree, it falls like a stone, with no sign of stability or control. When the flying snake approaches its target, it adopts a head-up posture and touches down tailfirst, perhaps 30 yards away from its launching position. Such snakes usually land in the branches of another tree, where they immediately resume their normal cylindrical shape.

From the same part of the world come the flying lizards, which, although they do not actually fly, come closer to having wings than do the frogs or snakes. The flying lizard's ribs support a membrane that runs along both sides of its body. Normally, this is folded and more or less hidden from view, but just before takeoff the ribs, "hinged" at their points of attachment to the vertebral column, are extended and the "wings" unfurled. On the neck, two flaps of skin seem to form accessory lifting devices, perhaps to support the heavier head region. Again, this small lizard can only glide, because the "wings" are incapable of providing the thrust necessary for powered flight, and merely help the little sky divers to plane down at a shallow angle from high vantage points.

The ace unpowered gliders are to be found among three groups of mammals: the flying rodents, the so-called flying lemurs, and the marsupial possums. All possess webs of furred skin on both sides of the body, stretched between the fore- and hindlimbs. As they leap into the air, they stretch their limbs sideways and turn themselves into flying carpets. In the flying squirrels and Australian possums the hands, feet, and tail are free, whereas in the flying lemurs (also known as colugos or cobegos) the extremities of the gliding membrane are supported by the fingers, toes, and a short tail. However, the flying squirrels' wrists have a backward-projecting spur of bone that supports the edge of the web along part of its outer side.

Their ability to alter the membrane tension and the set of their fluffy tails gives these gliders a large measure of control over their trajectories. In "flight," they can pitch, bank around in full circles, and, providing they hit a good updraft, even gain height. Landing—always a tricky maneuver for animals and aircraft alike—is accomplished not by an awkward belly flop, but by a neat four-point touchdown. On the final approach to a tree—gliding animals rarely land on the ground unless by accident—they erect the tail to a nearly vertical position, and this pitches up the nose, to bring all four feet forward into contact with the tree trunk. They are then likely to scamper upward in order to gain height, because the higher their launching point, the farther they can glide. Flying lemurs and marsupial possums can manage 100 yards or more between trees, sweeping down at a very shallow angle of 1 in 10.

Southeast Asia and Australia are the gliders' center of distribution. Of the 35 kinds of flying or gliding squirrel, one lives in North America, and one is widespread in Eurasia, but all the rest are found in the rain forests of the Southeast. They range in size from the 5½-inch pygmy flying squirrel, which, when spread out for gliding, would just about cover a small pocket handkerchief, to the doormat-sized giant flying squirrel,

with an overall length of four feet. In Malaysia and Borneo the squirrels share the canopy with two species of cat-sized flying lemurs, whereas in Australia their place is taken by five species of marsupials, from the mouse-sized feather-tailed glider to the three-foot-long greater glider.

The luxuriant rain forests that are the habitat of all these creatures have a continuous green canopy. But between the canopy and the ground there is an enormous area that can be exploited only if the problem of moving from trunk to trunk is solved. These three different types of mammal have independently come up with identical solutions to the problem of living in this middle space of the forest. Gliding is a means of extending the powers of leaping, and the gliders have evolved similar designs for diving through the air—an example of convergent evolution, whereby unrelated animals adopt similar "solutions" to common "problems."

The zoologist, in trying to comprehend how the gliding membranes must have evolved in the first place, assumes that "flying" mammals developed from nonfliers by gradually extending a web of skin between the limbs. Although the web is now adequate for allowing the animals to take great shallow leaps, it must have been too small during the early stages of its evolution to break their fall. And yet the flap of skin must have had

Gliding to earth requires a large drag-inducing surface area in proportion to weight. Because his surface area is relatively small, a sky diver (below left) plummets at a fatal speed until he artificially increases his surface area by opening a canopy such as this sport parachute (right). So-called flying mammals such as the flying squirrel (below) carry a built-in parachute in the form of skin flaps between the limbs. By stretching its limbs, the creature spreads its parachute, and thus glides safely downward.

Certain reptiles and amphibians have evolved into natural gliders, but in entirely different ways. This artist's impression of a flying frog (left) reveals enormously long, webbed toes that act as brakes to slow its fall to earth. The flying lizard (below) relies on folds of skin as "wings." By spreading its extra-long ribs, the lizard opens out the folds, rather as we open an umbrella. Flying frogs and lizards just glide from tree to tree. They do not fly. Both these strange kinds of animal occur in the Malaysian Archipelago, where they are well-designed for tropical forest life.

definite survival value. It is a commonly accepted theory that originally the loose fold of skin down the flanks may have served as a kind of protective camouflage. Even today the folded flap hangs down onto the tree trunk, eliminating telltale body shadows, and it is often colored to match the prevailing color of the bark. Only later, after it had grown, did its function as a gliding membrane become all-important.

It is also reasonable to suppose that there is more to the making of an ace sky diver than the simple addition of a billowing cloak to a squirrel or a possum. A powerless glider must have good instruments, a keen sense of balance, and the ability to accelerate downward, upward, and sideways in the wind. And then there must be a tendency to leap into space. Observations of the behavior of gliding mammals indicate that the mother has to encourage her brood to make their first launches; if a youngster stalls and crash-lands, it is retrieved and taken aloft for another try. This bears out the theory that evolution must operate on behavior as well as on physical structures. There are good reasons why we generally equip our airports with abundant rest rooms, shops, restaurants, and other distractions. Such amenities help to quell the natural fear generated by the nervous systems of terrestrial creatures faced with the prospect of being projected through the air.

Powered flight has been achieved by four independent groups of animals, each with its own evolutionary runway to the sky. Insects, first of all, became airborne nearly 400 million years ago—way ahead of the three kinds of vertebrate aeronaut. Of these, such reptiles as the pterodactyls were first into the air. Next came the birds, and with the development of their phenomenally light but strong feathers, they proved to have more survival power than the skin-winged pterodactyls. Finally the bats, representing the mammalian answer to birds, took to the air some 60 million years ago. The obvious characteristic of all the true fliers is wings,

Sailplane over the Texas coast. Man-made gliders exploit the same principles as the flying animals. Broad, tapered wings and horizontal stabilizers help to keep them up in the air.

and it is possible to see how each of the four kinds of aeronaut evolved these light, broad blades for summoning support from the air.

The one or two pairs of wings with which most insects are equipped are membranous extensions of the cuticle. In fact, an insect's wing is an object lesson in how to achieve immense strength over large areas with little material. A fly's wings, for instance, account for less than one per cent of the creature's total weight. Yet the membranes of which each wing is composed form an incredibly tough, stressed skin-structure, strengthened by a system of tubular spars that fan out from the base. In form the insect wing is not too dissimilar from that of an airplane, but it is unlike the fixed wing of a plane in that it beats up and down. In many insects, indeed—the hoverfly for example—the wing flips over during each beating cycle; this imposes a severe test on the wing structure and its materials.

In a recent experiment to assess the stamina of an insect, a fly was glued to a revolving instrument that forced it to remain in motion. As it flew, it was fed a sugar solution, and it managed to buzz non-stop for six days, covering nearly 250 miles, after which its tattered wings brought it to a stop. They had beaten over 88 million times—a formidable amount of flexing for such frail-looking blades.

Insect wings are not modified legs, whereas those of the vertebrate aeronauts are basically fashioned from the limbs. Bats, like the extinct pterodactyls, have thin skin wings stretched between the body, the limbs, and all or part of the tail. *Chiroptera*, the scientific name for the order that comprises the bats, means literally "hand wings," and bats actually support their highly elastic wings with four splayed fingers. This arrangement gives these flying mammals great control over the spread of the wings, and it is also a kind of insurance against total failure through breakage of a spar or ripping of the membrane.

Bats' legs are not of much use as an undercarriage, however, because the legs must support the membranous wings at the rear and must help to hold down the hind edges in slow flight. The extinct reptilian aeronauts probably labored under a similar handicap. From the engineering point of view, both bats and pterodactyls appear to have become structurally committed to flying at the expense of terrestrial agility. Both may well have gained their wings by way of the skydiving route that is still followed by the gliding rodents and marsupial possums.

The bat, then, is not nearly so well equipped as the bird for a terrestrial life. The bird is a dual-purpose creature of both earth and air, for its physical design developed from the design of its lizardlike ancestors, whose forelimbs were completely free to evolve into wings without involving the back legs. Anyone who has carved a chicken should be aware of the fact that the great chunks of meat on the drumsticks and breast are the "motors" that drive the legs and wings respectively, and the operation of the one does not interfere with the functioning of the other. This flexibility of design may have given the birds a competitive edge over the flying reptiles and may thus have contributed to the reptiles' downfall. The bird's wing also has great strength, because the whole arm, lengthened and fused wrist, and stout second digit form a stiff, and yet light, spar. Many of the bones in the wing are hollow, but are reinforced internally by a meshwork of trusses that give the wing extra stiffness without imposing a heavy burden of weight. When modern engineers construct bridges and aircraft, they try in similar fashion to provide maximum rigidity with the minimum material.

The feather, a unique "invention" of birds, must have played a significant part in making them such successful aeronauts. There can be little doubt that the evolution of feathers predated flight by millions of years. Originally, they might have been useful as a means of conserving body heat, for fluffy feathers are excellent insulators. But they have other virtues, too: they are lightweight, tough, and springy—in other words, wonderfully suitable for flight. Much of the surface of a bird's wing consists of especially elongated flight feathers that are inserted into the trailing edge of the arm. Each flight feather is constructed to withstand the forces of beating against the air and transmitting these forces to the leading edge spar.

Like all feathers, the wing feathers are made of keratin, a protein that also forms the dead outer layers of human skin; the central shaft of each quill is tubular in places, square in others, often filled with keratin foam for support, and has thickened ridges to allow distortion. If you have ever handled a single quill, you will appreciate the amount of bending and twisting that the structure can take without breaking—and yet a

A dragonfly's wings may seem the flimsiest part of the insect's body. Yet they are tough and strong enough to bear its weight and to speed the creature on its flight through the air.

feather weighs the merest fraction of an ounce. The vanes attached to either side of the central shaft are also a marvel in natural engineering, for they are composed of thousands of fine barbs exquisitely hooked together. On each wing there are 20 or more quill feathers, which overlap to form a nearly airtight blade. And these feathers are both smooth and elastic—smooth so that the wing can be folded and extended easily, and elastic so that when loads are removed the pinions regain their former shape.

If the vanes become unhooked, preening hooks them up again. But severe damage by breakage or from being eaten away by parasites can be rectified only by renewal, because feathers, like the outside of our skin, are dead matter. That is why birds need to molt their plumage once or twice a year. The molting process takes place gradually over the course of several weeks, and imposes a heavy drain on a bird's resources.

However, because the working surface of each wing is composed of many elements, the loss of one or two at a time—as happens with the majority of birds—during the molt does not significantly alter the aerodynamic performance. In this respect the birds are better off than the insects, which have nonrenewable dead wings, or than the bats, whose flying ability can be permanently impaired if their living skin wings are badly damaged.

Flying animals have neatly divided up the biological air space among them, with each type tending toward some kind of specialization. For example, insects are the smallest and slowest fliers and occupy the smallest niches. These vary in size as the insects themselves do. A microscopic wasp hardly more than one hundredth of an inch long can utilize the air space beneath a leaf or tiny toadstool, whereas a mighty atlas moth needs a forest glade to flop across. In general, birds take over on the size scale where insects leave off. The smallest hummingbirds are remarkably similar in design and size to the largest hawkmoths and may even compete with them for nectar, but the species at the top of the scale, such as the Andean condor and the wandering albatross, which have wingspans of 10 to 12 feet, inhabit the airspace over mountains or oceans.

Some insects do all or most of their flying by day, others at night. With the exception of a very few nocturnal groups—owls and nightjars, for example—birds operate by day, leaving the night sky to the bats. As the birds begin to settle in their roosts, a host of bats takes over the role of insect hunters, nectar drinkers, blood tappers, fruit eaters, and even fish catchers. The food opportunities open to creatures that can fly can be judged from the fact that there are at least 750,000 kinds of flying insects—the largest of all groups of related animals. Even among the vertebrates, the aeronauts rank high in the table of success if sheer number of flourishing species is taken as a valid criterion. There are 8580 kinds

Bats and birds are designed for flight in different ways. For instance, this Madagascar fruit bat (above) supports its wing membranes on modified fingers and legs; the legs are useless for walking. The sandhill crane (right) supports wing feathers on modified arms, and its long legs are left free for walking. It is thus a creature of earth and air.

of bird alone; and no less than one out of every four species of mammal is some type of bat. As an order containing about 1000 species, the *Chiroptera* come second only to the rodents.

What is the secret of flight? As we all know, there is much more to it than just waving wings around, for the wings must be shaped and moved in such a way as to summon powerful forces from flowing air. It is illuminating to draw comparisons between flying animals and airplanes, not only because of the many similarities but also because of some rather obvious differences.

A flying creature (whether an animal or a machine) needs to generate two forces acting in different directions. First it must produce a propulsive force or thrust. Aircraft create this by means of one or more rapidly rotating propellers or turbojet engines, which fling air backward. As the amount of thrust generated depends upon the mass of air and the speed by which it is shifted backward, and because air is very lightweight, a flying animal or machine has to move very large volumes of the atmosphere in order to obtain a useful motive force. Secondly, the downward pull of gravity can be defied only by means of an upward-acting lift force. In an aircraft, this is the role of the wings, which are shaped so as to generate powerful upward forces when dragged through the air.

Many people assume that a wing produces lift because a stream of air hits its undersurface and "floats" it up. In fact, the real force derives from its flattened tear-drop section, which causes air to accelerate over the curved upper surface. The speeded-up airflow causes a decrease in pressure on top of the wing relative to the straighter undersurface, and the wing is therefore "sucked" upward. This is the principle of the airfoil made use of in airplane wings. Increased curvature of the upper side of the wing, its angle of attack to the airstream, and the speed of the airflow all help to boost the amount of lift as long as air can be kept flowing smoothly around the wing. But if the flow starts to break up into

turbulent eddies, particularly over the upper surface, the lift is destroyed, and the wing is said to "stall."

The danger of stalling is greatest at low speeds—when, for instance, an aircraft is on the final approaches to landing, or just after lift-off, when the airspeed is low. To maintain as much lift as possible in such situations, the angle of attack of the wing is increased. Aero-engineers have also devised all kinds of high-lift devices to delay the stalling point, such as forewing slots, which, when extended, let through a stream of air to keep the sluggish boundary layer moving over the top of the wing. Extra lift can also be generated at low speeds by means of rear wing flaps, which can be used for increasing the wing area. When the flaps are fully lowered, they also help to delay the onset of turbulent flow by sucking air downward, thus holding the airflow close to the surface.

Living aeronauts differ from aircraft in that their wings have to take on the role of both the propeller (or jet engine) and the lifting device. The insects overcome this problem in much the same way as helicopters do. Helicopters manage to produce both lift and thrust from one structure—the rotor—by sucking air in from above and flinging it downward to produce a mighty draft on which the helicopter can ride. By tipping the rotor slightly forward, the pilot deflects the thrust backward in order to move along at a respectable speed. Insects' wings, although more like propellers than rotors, produce both lift and thrust in the manner of helicopters. But unlike the propellers of flying machines, which rotate constantly in one direction, the insects' wings move basically up and down, or occasionally forward and backward, in a reciprocating figure-of-eight motion.

This constant to-and-fro motion would not work if the pitch of the wing blades were fixed, for the thrust achieved on one stroke would be canceled out by a reverse thrust on the other. However, the pitch or angle of attack of the wing is altered as the stroke changes, so that positive thrust is achieved throughout the whole of the beat cycle. In fact, in such level fliers as the locust, the wings are not beaten in an absolutely vertical plane, but are tipped slightly backward; as a result, the thrust generated by the flexible propellers is deflected obliquely downward, and this keeps the insect airborne and moving.

Some insects, such as the hawkmoths, have perfected hovering as well as horizontal flight. They can do this by bringing their bodies into a vertical position, so that the thrust generated by the wings is used entirely for supporting their weight. Hoverflies keep their bodies horizontal and beat their wings in a horizontal plane. On the recovery stroke, the blades flip over along special lines of weakness, to maintain a steady downward thrust. A slight alteration of the stroke plane allows the insect to dart off in a flash. A similar approach to hovering is that of the hummingbirds and sunbirds, whose wings are rather stiff and bladelike. The amount of rotation possible at the shoulder joint of a hummingbird is so great that the wingblade can reverse its angle of attack on the recovery stroke—something that very few other birds can manage.

It takes an enormous expenditure of energy to support bulk by simply flinging air around. Bats and large birds have improved on this method, and they are much more comparable than the insects and hummingbirds to fixed-wing aircraft in that they utilize aerodynamic forces generated by airfoils to keep them airborne. The inner and outer regions of a bird's wing appear to have separate functions equivalent to those of an airplane's airfoil and propeller. During flapping flight, the outer part of the wing is moved

Fanciful 1893 design for a helicopter. In man-made helicopters, lift and thrust both come from spinning rotorblades. In insects, they come from one or two pairs of wings that beat up and down.

Alberto Santos-Dumont's "box-kite" aircraft accomplished Europe's first successful heavier-than-air flight at St. Cloud in France in 1906. By combining lift and thrust to become and remain airborne, man was beginning to discover the hidden ingredients of flight that evolutionary processes had built into certain insects, mammals, and those ace aeronauts the birds many millions of years ago.

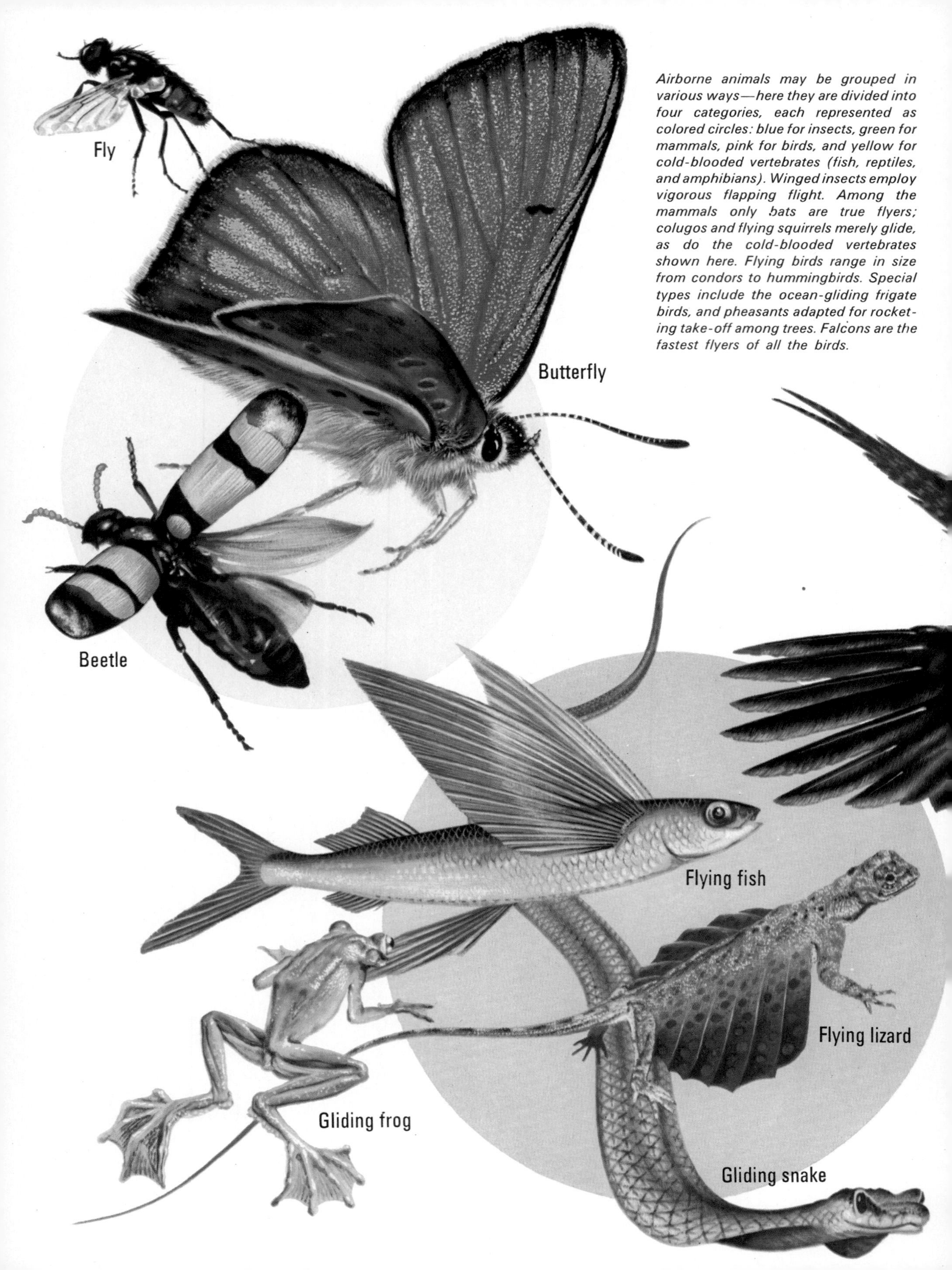

Airborne animals may be grouped in various ways—here they are divided into four categories, each represented as colored circles: blue for insects, green for mammals, pink for birds, and yellow for cold-blooded vertebrates (fish, reptiles, and amphibians). Winged insects employ vigorous flapping flight. Among the mammals only bats are true flyers; colugos and flying squirrels merely glide, as do the cold-blooded vertebrates shown here. Flying birds range in size from condors to hummingbirds. Special types include the ocean-gliding frigate birds, and pheasants adapted for rocket-ing take-off among trees. Falcons are the fastest flyers of all the birds.

LIFE TAKES TO THE AIR
Flying squirrel
Colugo
Bat
Pheasant
Andean condor
Frigate bird
Hummingbird
Falcon

obliquely forward and backward, and the pressure of air on the outer feathers inclines them forward, so that they bite the air and thrust it back, as a propeller does. On the recovery stroke, the flight feathers tend to separate, letting air through and thus reducing the resistance of the wing on its upward movement. Most of the lift is generated by the airflow over the inner wing, which is shaped like a typical airfoil and remains comparatively still throughout the flapping cycle.

Like aircraft, birds are equipped with accessory lifting devices and control surfaces, to keep the wings working as long as possible, particularly when the flier is about to land. The flexibility of a bird's wing, however, is far superior to the fixed form of an airplane's. By folding or stretching its wings, a bird can alter their surface area, and thus their lift, to suit changing circumstances. During landing and takeoff, the fully expanded wing can generate maximum lift, and the tail—which corresponds to the flaps of a plane—can add significantly to the amount of lift by being fanned and depressed. On the landing approach, the tail is invariably fully fanned; this helps to suck air over the inner surface of the wing, so that it can continue to work prior to touchdown, when there is a tendency for the smooth boundary layer of air to peel off into eddies. Some birds—frigates, swallows, and terns, for instance—have streamers that lie close to the trailing edge of the wing when the tail is spread out, and these may form slots to help the wing to work better close to the stalling limit.

Even its feet may help a bird to fly. Such web-footed creatures as the small-winged auks often use their expanded feet as auxiliary lifting devices. Others use the feet as air brakes. This is particularly characteristic of gulls as they come down to land or try to sink down against the violent uprush of air that often occurs near cliff edges. Vultures, with their high-lift wings, lower their legs in order to decrease their speed when swooping down to a carcass. The white-backed vulture even has a small web between its third and fourth toes, presumably to increase the effectiveness of the foot as an air brake.

A trident airliner (above) and an African vulture (below) both avoid stalling when traveling at low speed; the 'plane by lowering its wing flaps, the bird by lowering and spreading its tail. Vultures fly largely by gliding, stiff-winged, on rising currents of warm air. This stresses their apparent similarity to man-made airplanes.

Flying is of little use, needless to say, unless the aeronaut can achieve a large measure of control over where he goes. It is in their solutions to this problem that flying animals and aircraft differ fundamentally. An airplane is designed to be flown by relatively slow-reacting earthbound creatures who have a poor sense of movement in three-dimensional space. Stability and safety are of paramount importance, and so the machine must move through the air on a predictable course, with very few control adjustments. If a flying animal had this inherent tendency to fly straight, like a paper dart, its chances of survival would be slim. When chasing prey or escaping from an airborne enemy, maneuverability is what counts, and maneuverability and stability are mutually exclusive.

It should not surprise anyone familiar with the evolutionary process that "prototype" flying reptiles and birds were more stable fliers than later forms. The primitive bird that we call *Archaeopteryx* had a long, thin tail with feathers along either side, and one kind of pterodactyl had a tail that terminated in a fleshy tab. There can be little doubt that these appendages functioned like the streamers on a kite and helped the early aeronauts to fly true, with a minimum of second-by-second adjustments. With the passing of time, the brain and sense organs became sophisticated enough to keep control without the help of such contrivances. Accordingly, tails became shorter, and both birds and pterodactyls became increasingly capable of maneuvering. (A similar refinement has taken place in combat aircraft design; in the most maneuverable airplanes, a tolerable degree of instability has been achieved by handing over the detailed control of pitching, swaying, and rolling to a quick-acting computer.)

The tail of a modern bird is in no way analogous to the tail plane of an aircraft. It is equivalent to a rear wing flap, and birds that have their tail feathers experimentally removed can fly tolerably well, although they may have to land at rather high speeds. To fly successfully, an animal aeronaut must obtain enough information from its sense organs to be able to assess what its body is doing at any one instant, must compare this with what it should be doing, and must then program its control surfaces to take necessary corrective measures.

The split-second adjustments that flying ani-

Left: a landing seagull's bent, angled wings, fanned tail, and outthrust "air-brake" feet are all control surfaces that a highly maneuverable flying bird needs. Maneuverability aids survival among animals of the air. Right: artist's reconstruction of an Archaeopteryx, *the earliest known bird, which first flew or glided about 140 million years ago. A long, feathered tail ensured stability, but* Archaeopteryx *was less maneuverable than modern, short-tailed birds.*

mals must make in order to stay airborne are performed mainly by the wings. Insects can alter the direction of thrust by slight changes in the wings' stroke plane, and the pitch of their "propellers" is also adjusted by special muscles in the thorax. As for steering—well, a fly can spin on the spot simply by shutting down one wing for an instant while maintaining full thrust on the other. Beetles, on the other hand, cannot stop a single wing from beating, but they can vary the thrust of the two wings independently by decreasing or increasing the arc through which each wing beats. But the birds, with their highly flexible wings, are the real masters of controlled flight. And some of them are large and slow enough for us to be able to observe with the unaided eye how they do it.

It is an exhilarating sight to watch gulls constantly adjusting their trim as they soar over cliffs. To hold a steady course, they reduce or increase the lift of a wing by either folding or extending it. In this way, too, they keep from rolling. And they counteract pitching beak-down or beak-up by moving their wings fore or aft, using the center of lift to trim their bodies one way or the other. Furthermore, the use of their feet as air brakes controls swaying and yawing.

Flying requires a great deal of power as well as control, and all flying animals have massively developed muscles for moving their wings. A bird's power plant accounts for up to one third of its body weight. The flight muscles are mostly located on a keel-like extension of the breastbone. In fact, the breast flesh consists of two pairs of muscles, one pair on each side of the breastbone. In each pair, one muscle lies beneath the other. The superficial one (which is the first to be carved on a roast chicken) is by far the larger; when it contracts, it pulls the wing downward on the power stroke. The smaller and weaker muscle beneath is capable of raising the wing by virtue of the fact that the ligament from it runs up over the shoulder and into the *humerus* (the long bone that corresponds to a human being's upper arm bone). So the chief flight muscles are neatly slung beneath the center of lift of the wings to preserve the bird's balance during flight. So powerful are these muscles that the whole of the chest is strengthened by a stout compression strut, to prevent their combined pull from collapsing the bird's body.

Bats have a functionally similar arrangement, except that they lack the birds' keel-like extension of the breastbone. Instead of being attached to this, the chest muscles on each side are joined mid-line by a band of connective tissue, and so the two pairs pull against each other.

The flight "motors" of insects are crammed into the *thorax*—the middle section, between the head and the abdomen—where they occupy about 85 per cent of the space. A fly's motive forces are produced by no fewer than 42 different muscles

The tube-nosed bat (left) and the hummingbird (right) conserve energy by lowering their temperature while sleeping. On waking, they perform a pre-flight warm-up to raise their body temperature to the level at which the flight muscles operate.

controlling the power and pitch of the wings. In the case of an insect that flaps its wings slowly, such as the locust, the muscles are anchored to the base of the thorax, with their other ends inserted into the wing roots, so that the muscles are able to move the wings directly as soon as they are stimulated by the motor nerves. But nerves need a little time for recharging between passing messages, and so such insects as gnats, which beat their wings at up to 1000 times each second, are forced to twitch their muscles at a much faster rate than that at which nerves can relay stimuli. They do this by means of enormously complicated mechanisms that control the flight muscles within the thorax. Such mechanisms make the arthropods astoundingly efficient denizens of the air.

Their flight muscles are certainly among the most active of living tissues, and they burn up prodigious amounts of fuel in order to exert tremendous power in relation to weight. For example, the output of a honeybee's "motor" is much the same, weight for weight, as that of an early aircraft's piston engine—nearly one horsepower per second. In comparison, a human leg muscle is a very poor performer indeed; it can just about manage 0.05 horsepower per pound. Most flying animals can muster a great deal of extra power over and above their resting requirements. A budgerigar in flight expends power up to 20 times faster than its resting rate—a performance that a person of exceptional athletic ability would have a hard time duplicating for more than a few seconds. And a May beetle (or June bug) can manage a staggering increase in peak flying power of 107 times its resting rate.

One characteristic that enables flying animals to summon forth such a sustained output of energy is their high rate of metabolism. For example, birds have a body temperature of at least 106°F. At that level of heat, the chemical reactions within the body take place quickly and fuel is burned rapidly. It is not difficult for most birds to maintain such a high temperature, because they are covered by an insulating layer of fat and feathers. Nor is it hard for the day-flying insects, some of which are warmed by the heat of the sun, others of which—such as the big bumblebees—are warmly clothed in a dense pile of "hair."

Animals that fly in the cool night air have a problem. Large insects such as hawkmoths, which cannot get sufficient power from their flight muscles until they reach an internal temperature of between 95°F and 100°F, must go through a wing-twitching routine before takeoff, in order to warm up the muscles. By burning up fat, the massive twitching muscles generate heat within the thorax, and this warming-up process may last for several minutes before the creature

has worked up sufficient power for a lift-off.

Bats and hummingbirds also go through a preflight warm-up after having cooled down during their sleeping period. There is an economic advantage in allowing the body to cool while resting. To maintain the high temperature essential for operating powerful flight muscles requires a very great consumption of fuel, because the hotter the body, the greater the loss of heat to the environment. Very small insects lose heat to the surroundings so quickly that their only practical means of achieving a high temperature is to pick it up directly by basking in the sun. Larger aeronauts, such as most birds, find it comparatively easy to hold on to body heat because they have small surface areas in relation to bulk. In between these two extremes—among the largest insects, smallest birds, and most bats—the heat profit-and-loss account favors alternate cooling and heating up.

Once airborne, the situation changes, for the highly tuned flight "motors" produce so much heat when hard at work that overheating becomes a danger. When a moth's temperature

Ornithologists reckon that no birds much heavier than about 25 pounds can develop the muscle power required for flight. Three heavyweight fliers appear on these pages. To become airborne, the white pelican (left) and the trumpeter swan (center) pick up airspeed by taxiing across a sheet of open water. The California condor (right) prefers to flop off a steep slope into a rising air current. Condors can cover immense distances by simply soaring through the air on their long, broad wings using little muscle power.

soars to around 105°F, its thorax muscles are cooled by the blood, which transfers surplus heat to the abdomen, from which it passes into the air. The bodies of large beetles and locusts are porous enough to allow a little moisture to seep out, and so they lose heat by evaporative cooling. Flying birds are probably kept cool by the air sacs distributed throughout their bodies. And bats have a very efficient air cooler in the form of their vascular wing membranes, for these can be flushed with blood, which then dumps the excess heat into the airstream. The pterodactyls were probably also air-cooled in this manner.

Such powerful living engines need high-capacity pumps and plumbing to keep them supplied with fuel and oxygen and to rid them of waste products. Birds and bats have bigger hearts than those of earthbound creatures of similar size; for example, a sparrow's heart is 2.7 times larger than that of a mouse, and the heart accounts for about 1 per cent of a bat's full body weight as opposed to about 0.5 per cent for a small nonflying rodent. And birds have unusually complex respiratory tracts, whose special construction must surely be related to the need for an enhanced intake of oxygen and output of carbon dioxide.

The sturdy flight muscles of insects are threaded by a number of air pipes, which bring in vast quantities of air. In fact, the air pipes do

Rhamphorhynchus *(left) and royal albatross (below)—ancient and modern types of animal designed for flight.* Rhamphorhynchus *belonged to a group of flying reptiles that evolved parallel with the pterodactyls. It had a skin-membrane wing stretched between an elongated fourth finger and its knee, and was capable of powerful flapping flight. Possibly, climatic changes brought rough weather that wiped out these flying reptiles. In contrast, the albatross needs strong, constant winds to keep it airborne, as it lifts and dips majestically above the ocean waves.*

double duty as both inlet and exhaust manifolds, taking away carbon dioxide as well as transporting oxygen to individual muscle cells. The entire thoracic tracheal system is ventilated by the pumping action of both the flight muscles and special respiratory muscles, which squeeze the air tubes intermittently, forcing the air to circulate. Some remarkably high flow rates have been found to occur in butterflies; during a 10-hour flight, more than one gallon of air can flow through their hair-thin tracheae. That much air must be equivalent to a formidable gale passing through the tiny tubes.

The power developed by an animal's flight "motors" has an important bearing upon the maximum size it can attain. Obviously, hummingbirds could not grow to the size of vultures and still function as hummingbirds, nor could diminutive vultures soar at will in rising currents of warm air. In other words, large animals are not simply scaled-up versions of smaller ones; fundamental changes in design must take place in line with the laws of engineering. For example, very small wings produce hardly any lift in a windstream because the air tends to stick to the small surfaces. That is why insects and small birds devote most of their effort to getting the air to move by beating their wings very rapidly. Conversely, large wings can behave like conventional airfoils and generate lift, so gliding on motionless wings is more easily accomplished by storks than by sparrows.

There are other considerations, too. With an increase in size, weight increases at a proportionally greater rate, and so the heavy aeronauts need relatively immense wing areas to support their bodies. In practice, large fliers must also fly faster, in order to generate sufficient force to keep their heavy bodies airborne. Naturally, greater speed can be gained only by the development of more powerful engines. Unfortunately, muscle power increases at a much slower rate than weight. Thus, as flying animals grow bigger, they get relatively weaker. To be precise, if a bird doubles its weight, it needs 2.25 times as much power to stay airborne, but its muscles can deliver only 1.59 times the power. Clearly, the power margin available to a flying animal drastically declines as it puts on weight.

Insects are phenomenally strong for their size, and some can carry enormous payloads. An ichneumon fly, for instance, can carry a caterpillar several times its own weight through the

Ménuires
14

air in order to provision its nest. The most energetic kind of flying is hovering, for it demands a sustained thrust equivalent to the weight of the hoverer—and this is well within the capability of the insects, as it is of such very small birds as the hummingbirds and sunbirds. But it becomes increasingly difficult for larger birds. Pigeons can sustain hovering flight at maximum power for just a second or two, but vultures have no chance at all and can barely fly level in still air.

A bird incapable of generating enough full-throttle power for hovering is also incapable of taking off by simply jumping into the air. Instead, it must taxi along the ground or water, as a man-made plane does, or else it must fall off a high vantage point in order to pick up sufficient air speed to get the wings working—which, indeed, is what most heavy birds do. There comes a point, too, at which the power required to keep a heavyweight in level flight can barely be mustered by the flight muscles. For birds, the maximum possible weight for flight has been calculated at about 25 pounds, and this corresponds closely to the size attained by such fliers as the mute swans, kori bustards, white pelicans, and California condors.

Some animal aeronauts have managed to break through this weight barrier, but they are now mostly extinct. And, like many of the larger birds living today, they had to depend substantially on the power of the atmosphere as well as on their own muscle power to keep them aloft. Thus we should call them "soarers" rather than "fliers."

Two fundamentally different glider designs have been evolved by living aeronauts (as opposed to the much more restricted group of sky divers mentioned at the beginning of this chapter). One type of soaring creature manages to remain airborne by means of its long, narrow wings; the other has broad, arched wings that have been expanded out of all proportion to body weight. The long, narrow wings are characteristic of such oceanic gliders as the albatrosses and frigate birds. Their slim wings can slip through the air without producing high drag

Man's muscles are too weak to be able to power artificial wings in flapping flight. But inventors have produced successful fixed-wing "human kites." Below: Otto Lilienthal, a German aeronautical engineer, made numerous flights from a man-made hill near Berlin before dying when his glider crashed in 1896. Left: a modern kite flier slope soars like a gull after leaping into the air from a high cliff.

White pelicans in flapping flight remind us of the relatively vast energy output required by this activity. The highly efficient fuel for long-distance flights is fat, which is stored in a bird's body.

forces, and so they are nicely adapted to operating in the high winds that blow over the sea.

These birds make use of the fact that the wind is slowed down by friction close to the waves, and reaches its highest speed at 50 to 100 feet up. Albatrosses wheel around in this upper layer of air, picking up speed in the fast windstream, gliding down very swiftly at a shallow angle, then heading into the wind at low level and using the momentum to sweep them upward once again. There are, however, limitations to long, narrow wings. Because of their high inertia, it takes outsize quantities of energy to get them in motion, and in the absence of brisk winds the birds are easily fatigued and find it difficult to take off. So it is not suprising that albatrosses are mostly confined to regions where the westerly winds blow. There is also a structural limitation on the maximum span of the narrow wings. As they become longer, the upward leverage exerted on their undersides escalates. Too much length would necessitate a heavier bone support than could be carried through the air by the birds.

The alternative glider design—a disproportionate expansion of the wing area—was undoubtedly what happened with the pterodactyls, which were extremely lightweight creatures as compared with birds of similar body size. The very summit in broad-wing, lightweight-body construction was reached by the pteranodon, a superoceanic gliding reptile with a wingspan of up to 26 feet. If there were a bird with such a wingspan, it would have to weigh about 200 pounds, and it would be far too heavy to fly under its own power. But the pteranodon tipped the scales at between 35 and 40 pounds—equivalent to the amount of baggage that a passenger may carry free on a modern airliner.

This incredible flying dragon had underdeveloped muscles and tubular inner-wing spars as wide as a human arm bone, but with walls no thicker than blotting paper. It could evidently glide at speeds as low as 12 miles an hour on very light winds (most birds would fall to earth if they attempted such a feat). However, its frailty made it ill-equipped for rough weather. Climatic changes bringing strong oceanic winds may have led to the downfall of this magnificent flying reptile, which was already extinct by the end of the calm, tropical Cretaceous period, more than 60 million years ago.

Land soarers are quite different in shape from the elegant, thin-winged, open-sea soarers. Although long, narrow wings give the best gliding performance, the design of such birds as the vulture has been greatly modified by two considerations. First, the bird must be able to take off in still air from ground level perhaps several times a day—and with a heavy load of food in its crop. Because long wings are difficult to flap, and flapping is a necessity for such take-

offs, natural selection has reduced the wingspan, compensating for the loss of length by an increase in width. Secondly, because the upward-moving masses of warm air are sometimes very narrow, the soaring bird needs to be able to turn in tight circles within confined airspaces. The high-speed, narrow-winged gliders cannot do this; they require wide columns of air in which to maneuver. But the broad, highly arched wings of such birds as eagles, storks, pelicans, and vultures give them a distinct advantage in the narrowest columns of rising air.

With such a broad, compact surface area to carry the weight of the body, such birds have comparatively slow sinking speeds in still air. As long as they can find a column of air that is rising faster than their sinking speed, they can gain altitude. When emerging from the top of a current of warm air perhaps several thousand feet above the ground, they can glide for eight or nine miles, losing height slowly, until they find another such thermal. As a means of traveling and searching for food, this is the only economical method of cross-country flying for birds that are approaching the size limit of self-powered flight. For example, vultures must spend a great deal of their time in the air scanning the ground for food, and even when nesting must often travel 70 or 80 miles away from base in search of prey. Only through using rising columns of sun-warmed air can these birds hope to make a living. Similar soaring techniques have evolved in a number of unrelated species that live in areas of strong year-round thermal activity, once again providing a case of evolutionary convergence.

The only major drawbacks of cross-country soaring are that the big birds can move only by day when the sun has warmed the land sufficiently to create updrafts, and they must skirt around large open stretches of water. But for temperate-zone migrants the size of white storks, cranes, and eagles, it is a very economical way to travel, for they may be able to complete their annual long-distance flights on one thirtieth of the power necessary for flapping flight.

Long-distance flights in general need good supplies of fuel, and small flying animals can eat relatively enormous amounts of food because of the favorable power-to-weight ratio. The natural equivalents of the petroleum products that power man-made aircraft are sugars and fats. Of the two, fat is the more efficient fuel. The trouble with sugars is that they are soluble, and so a flying animal that burns them can carry only as much as its body fluids will tolerate in solution. In practice, this is not much, and so such sugar burners as flies, bees, and hummingbirds need to stop and refuel very often.

The super-grade fuel used by long-distance fliers is fat, which weight for weight has nearly as high an energy yield as gasoline. Because fat is not soluble in water, the aeronaut's body can store quite large quantities of it without upsetting the chemical balance of tissue fluids. When cruising on fat, both locusts and birds are likely to consume no more than about one per cent of their body weight per hour of flight; and because many small birds can nearly double their weight during the premigration period, this supply of fat will keep the flight "motor" running, without replenishment, for several days. Indeed, sedge warblers can fly from southern England, across France, the Mediterranean, and the Sahara Desert, to central Africa in one stage.

The sheer economics of flying may have severely limited the evolution of completely aerial animals. No aeronaut, no matter how splendidly adapted to life in the air, can spend all its time in flight. On the whole, most fliers subsist on the plant and animal life that covers the earth, and they are airborne for only a small percentage of their time. Many, however—among them, swifts and albatrosses—do spend very long periods on the wing searching for food. And although most seabirds can settle comfortably on the water between flights, a few cannot. The plumage of oceanic sooty terns and frigate birds is not well enough waterproofed to allow them to rest at sea, and so they must remain airborne for months at a time between breeding seasons.

The nearest approach to a totally aerial form must be the swifts, which feed on atmospheric "plankton" and can even sleep on the wing. Courtship among the swifts is an aerial affair, too, and couples can mate while carefully flying together, one above the other, laying their eggs in crevices high up on cliffs or buildings. Many insects also copulate on the wing.

Is there a chance that someday an inventor will make a pair of wings with which a human being can achieve flight by means of his own muscular power? It seems highly unlikely. At our weight, we should need to be equipped with wings the size of the pteranodon's (a 26-foot span) and breast muscles six feet deep. Even then we should require the help of gale-force winds to get us off the ground. As natural walkers and runners, we have pumps and plumbing systems strong enough to keep our leg muscles going, but not much more. Our body chemistry, in short, is not capable of yielding the surge of power necessary to support our weight in the thin, gaseous medium of the air.

If you sometimes feel a bit cheated because you cannot soar aloft on your own like a skylark or a nightingale, there should be comfort for you in the thought that although we human beings never evolved wings, we have made up for the lack by acquiring unique brain power. And so, after all, the abiding collective genius of our species *has* conquered the air. Even as compared with the birds we are rather remarkable, for we have fashioned our own metal creatures, which can fly hundreds of people many miles through the troposphere at any time in almost any weather.

Boeing 727 landing at Washington Airport, guided by approach lights (foreground). Airplanes, which carry hundreds of people thousands of miles and far faster than any bird can fly, compensate modern man for his own lack of built-in wings.

Index

Page numbers in *italics* refer to illustrations or captions to illustrations.

Picture Credits

Key to position of picture on page: (B) bottom, (C) center, (L) left, (R) right, (T) top; hence (BR) bottom right, (CL) center left, etc.

Cover: Dr. J. A. L. Cooke/Oxford Scientific Films
Title Page: Tony Stone Associates
Contents: A.G.E./Östman Agency
8–9 N.A.S.A.
11 N.A.S.A./Aspect
12 Spectrum Colour Library
13 Ronan Picture Library
14(T) Mary Evans Picture Library
14(B) Marc Riboud/Magnum
15 Brian Brake/The John Hillelson Agency
17 Spectrum Colour Library
18 Ronan Picture Library
19 A.G.E./Östman Agency
20 Photo Researchers Inc.
24 T. Lövgren/Östman Agency
25 G. R. Roberts, Nelson, New Zealand
26 Bill N. Kleeman/Tom Stack & Associates
28 Photo Researchers Inc.
30(T) Victor Englebert/Susan Griggs
30(B) Spectrum Colour Library
31 G. R. Roberts, Nelson, New Zealand
32 Van Bucher/Photo Researchers Inc.
33(R) Dr. Wm. M. Harlow/Photo Researchers Inc.
35 Spectrum Colour Library
37 Sue McCartney/Photo Researchers Inc.
38 Ron Church/Tom Stack & Associates
39(R) *Daily Telegraph* Colour Library
40(T) Fujifotos, Tokyo
40(B) Ian Berry/Magnum
41 Hallinan/FPG
43 Spectrum Colour Library
44–5 Tom Stack & Associates
46(T) Heather Angel
46(B) Eileen Tanson/Tom Stack & Associates
47 Robert B. Evans/Tom Stack & Associates
48(T) Al Giddings/Sea Library/Tom Stack & Associates
48(B) David Houston/Bruce Coleman Limited
49 C. Roessler/Sea Library/Tom Stack & Associates
51(T) G. Bradt/Frank W. Lane
50(B) Dr. C. C. Wendle/Tom Stack & Associates
51(B) W. M. Stephens/Tom Stack & Associates
52 Jim & Cathy Church/Sea Library/Tom Stack & Associates
53(R), 54 Heather Angel
56(L) Raymond A. Mendez/Animals Animals © 1974
56–7(C) Heather Angel
57(R) Kenneth Read/Tom Stack & Associates
59 Tom Myers/Tom Stack & Associates
60 D. & K. Urry/Bruce Coleman Ltd.
62 Tom Myers/Tom Stack & Associates
63 Photo Fred Bruemmer
64(L) Vienne Bel/Pitch
65(T) Warren Garst/Tom Stack & Associates
65(B) Ferrero/Pitch
66 Dr. E. R. Degginger/Tom Stack & Associates
67 Dr. H. Hood
68 Sven Gillsäter/Tiofoto
69 A. Jolly/Photo Researchers Inc.
70(L) Heather Angel
71 A–Z Botanical Collection
72 Earl Kubis/Tom Stack & Associates
73 Steinhart Aquarium/Tom McHugh/Photo Researchers Inc.
75 A. Joyce
76 de Bellabre/Pitch
77 G. R. Roberts, Nelson, New Zealand
78(B) Larry Moon/Tom Stack & Associates
79 Jacques Jangoux/Photo Researchers Inc.
81 Bill Leimbach
83 Paul A. Benson/Animals Animals © 1974
84(B) G. R. Roberts, Nelson, New Zealand
85 Alain Nogues/Sygma
86(T) A. Bailey/FPG
87 Dennis Hallinan/FPG
88(T) W. J. Vogel/Tom Stack & Associates
88(B), 89(TR) G. R. Roberts, Nelson, New Zealand
89(B) A.G.E./Östman Agency
91 Adam Woolfitt/Susan Griggs
94 Photo by Willis Wipf, Freeman, S. Dakota
95 Charlie Ott/Photo Researchers Inc.
96(L) Fred Baldwin/Photo Researchers Inc.
97 Tom Myers/Tom Stack & Associates
98 Heather Angel
99 Bruce Coleman Ltd.
100(L) A–Z Botanical Collection
100(R) G. R. Roberts, Nelson, New Zealand
101 A-Z Botanical Collection
102 George Holton/Photo Researchers Inc.
104(B) Susan Griggs
105 D. & J. Bartlett/Bruce Coleman Ltd.
106 D. & K. Urry/Bruce Coleman Ltd.
107 Philippa Scott/Photo Researchers Inc.
109 Shelley Grossman © Time Inc. 1975
110 Warren Garst/Tom Stack & Associates
111(R) Martha E. Reeves/Photo Researchers Inc.
112 Jeff Dixon/Photo Researchers Inc.
113(T) S. L. Waterman/Photo Researchers Inc.
113(B) William A. Welch/Photo Researchers Inc.
114 Mary Evans Picture Library
115 James A. Sugar/Photo Researchers Inc.
117 Bellieud/Pitch
118(L) Albert Moldvay/Photo Researchers Inc.
119 Jeff Foott/Bruce Coleman Inc.
120–1 Ronan Picture Library
124(T) British Airways
124(B) Montoya/Pitch
126 Adam Woolfitt/Susan Griggs
128 Eric Lindgren/Ardea, London
129 Courtesy of The American Museum of Natural History
130(L) Russ Kinne/Photo Researchers Inc.
130(R) Photo Fred Bruemmer
131 Tom McHugh/Photo Researchers Inc.
132(B) M. F. Soper/Bruce Coleman Inc.
134 Bruno Barbey/Magnum
135 The Bettmann Archive
137 Aldebert/Jacana
139 Bruce Frisch/Photo Researchers Inc.

Artist Credits

© Aldus Books: Susan Cook 103; Design Practitioners 127, 132(T); David Nockels 21, 23, 80, 92–3, 122–3

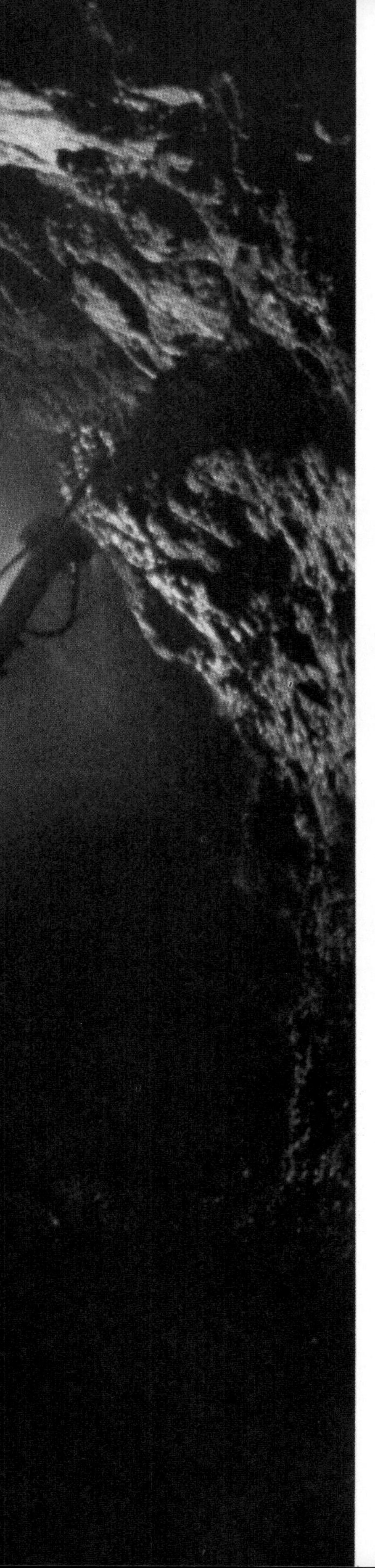

PLANET EARTH

Part 2
The Earth's Crust

by Arthur Bourne

Series Coordinator	Geoffrey Rogers
Art Director	Frank Fry
Design Consultant	Guenther Radtke
Editorial Consultant	David Lambert
Series Consultant	Malcolm Ross-Macdonald
Art Editor	Douglas Sneddon
Editor	Damian Grint
Copy Editor	Maureen Cartwright
Research	Peggy Jones
Art Assistant	Michael Turner

Contents: Part 2

Introduction

Of all the components making up our daily environment, the one we take for granted more than most is that thin crust of rocks lying beneath our feet. Yet it is this that protects us from instant annihilation, provides the necessities for life, and makes our modern technological civilization possible.

The story of the relationship between life and the crust goes back to the earliest days of the earth's creation. The crust was formed in those turbulent times, and as it cooled and the seas appeared, life began. The subsequent evolution of life is a story of a long progression of life-forms, a tale of the slow emergence of a complex intimate relationship between animals and plants, the atmosphere, the oceans, and the earth's crust. This system of relationships constitutes the world ecosystem, a system that is finely balanced by the energy of the sun on the one hand and by the energy generated beneath the crust on the other.

During the great span of earth history the planetary crust has undergone many changes. There were summer periods that lasted hundreds of millennia, when life flourished, and times of great cold, when only the most adaptable survived. One of the most successful animals at surviving was our own kind—man.

Today modern man dominates all other forms of life, a position attained through the utilization of the materials of the earth's crust. Yet strangely he ignores the changes he is bringing to the delicate but vital life-crust relationship. He forgets that in spite of all his ingenuity he is still dependent on that partnership.

In this book we explore the history and nature of that fantastic relationship between earth, life, and man. The story of the earth's crust is our story too.

The Moving Landscape

In a limitless universe, in a galaxy of stars among countless millions of galaxies, orbiting one star in a family of 100,000 million stars, is a small spinning world—an abode of life, the home of man.

Years roll by, generations are born and die, great civilizations rise and fall. Only the mountains, hills, valleys, and ocean coasts seem permanent. Protected by their planet from the hazards of the universe, men live out their short lives in comparative safety. The physical world appears to be a stable and safe place, and on the human time scale it is for the most part secure. Only when the land shudders with earthquakes and rumbles with erupting volcanoes do men ponder the violence that lies a few miles beneath their feet.

Although the earth is quieter now than in times past, powerful forces within its interior continue to build and shape the continents in a drama that is being played out continuously day by day, year by year, century by century, millennium by millennium. In even the most placid-looking landscape dynamic changes are taking place, but by their very nature they will not be completed within the lifetime of the onlooker. Hundreds of thousands, if not millions of years will pass before they are completed. Yet, in some far-distant tomorrow, the great mountains of today will be insignificant hills and the sediments of the seas will be high mountains.

The changes that occur throughout the earth's history are caused by the interplay of forces inside the planet and those that act upon its surface. Heat generated in the earth's interior by radioactivity alters the structure and distribution of the earth's crust, while the external forces of weathering and erosion tend to counter them. Between them these conflicting forces shape the external features of our planet.

The earth's crust, stretched like a skin over the globe, floats on the planet's inner layer. Compared to the planet as a whole the crust shrinks to insignificance, for nowhere is it much more

Water in liquid and solid forms has played a powerful role in sculpting the surface of the land. Here, in southeastern Utah, the San Juan River has gnawed a snaky canyon 1000 feet deep in the arid weather-beaten surface of the Colorado Plateau.

than 25 miles thick and under the oceans its depth is often less than four miles. On a globe such as the earth, some 8000 miles in diameter, these dimensions are inconsequential, for if we were to represent the earth by a sphere six feet in diameter, the crust would be no thicker than a layer of paint spread over it. On this scale even the highest mountains would hardly be detectable. Yet it is this thin skin of rock, particularly the part that makes up the continents, that we call home. With its oceans, landmasses, and enveloping atmosphere the crust forms the environment into which we and all living creatures are born and inextricably bound.

This beautiful small planet of ours is tied by gravity to an unspectacular star in the vastness of the universe but, despite their smallness and loneliness, the earth and the sun provide the conditions that make life possible. Many stars have a zone in the space around them in which a planet—if any exist there at all—has a good chance of developing and maintaining life. Known as *biotic zones*, these circumstellar regions vary in width and distance from their star according to its size and radiation. Although the earth is positioned approximately in the

Scenes from the polar regions remind us that much of the world looked like this in the cold spells of the Ice Age—times that could well come again. Left: mountains of the Antarctic Peninsula rise above ice that juts out into the Gerlache Strait. Below: shipboard view of an Antarctic survey team dwarfed by the ice-barrier upon which it has landed. Several such immensely thick slabs lie off Antarctica, products of glaciers that creep down to the ocean. Right: pack ice sighted in the Arctic Ocean near the North Pole. Much of the Arctic Ocean surrounding the North Pole wears a perpetual skin of floating ice.

middle of the sun's biotic zone—neither too near to be too hot nor too far to be too cold—life of the kind we know would not be possible without protection from the destructive ultraviolet radiation given off by the sun.

Only about half of the sun's radiant energy reaching the earth's atmosphere penetrates to sea level. Much of the sun's harmful ultraviolet radiation as well as other potentially lethal radiations from the far reaches of space are filtered off in the upper atmosphere, and clouds and dust particles reflect a considerable portion of the incoming radiation back into space. It is this reflected radiation that we call earthshine.

A further portion is held temporarily as heat in the atmosphere's water vapor, leaving about half to be absorbed as heat by the land and, most importantly, the oceans. A tiny but significant proportion of the sunlight—less than one tenth of one per cent—enters the cycle of life through the leaves of green plants. But without the oceans and the atmosphere—most particularly its water vapor and carbon dioxide, which trap heat—almost all the incoming energy would be immediately re-radiated into space.

While it is temporarily stored in the oceans and

atmosphere, the sun's energy drives the earth's complicated "heat engine." By human engineering standards this heat engine is inefficient but it is the powerhouse behind all the activity taking place on the earth's surface. It powers the great currents and waves of the seas, the movements of air masses in the atmosphere and the complex chemistry of life.

Eventually, however, almost all the radiation received from the sun must return to space, otherwise our planet would become a hot, lifeless world. On the other hand, the temporary holding of heat and its circulation ensures that the sunless side of the earth does not freeze up altogether. During the long span of the earth's history this equilibrium between heat gain and heat loss has, from time to time, been thrown out of balance and the living regimes have been threatened by extremes of heat and cold, but the equilibrium has so far always been restored in time to save life.

As the earth turns on its axis, alternate heating and cooling as the different regions turn into sunlight and out again into the shadow of night cause the atmosphere to expand and contract to produce a continual circulation over the entire planet. Warm air currents and direct solar radiation ensure the evaporation of water from the ocean surface and its circulation around the world. When these water-carrying air masses are cooled, water vapor in them condenses into its liquid form. When the droplets grow too heavy to be borne on the wind, gravity pulls them back to the earth as rain, snow, or hail. These are the first moves in what is a very complex system of water circulation within the environment—the water or *hydrological* cycle.

Water, the most important single factor in shaping the surface of the land, is a strange substance in that it exists in all its physical states—solid, liquid, and vapor—within the narrow range of temperatures and pressures found on the earth's surface. It is surprising that there is any liquid at all, for only one other is known to occur naturally in any abundance and that is petroleum. Although some of the earth's water is found as vapor in the atmosphere, by far the greatest bulk of this unique substance occurs

Aerial view of parallel mountain ridges in the European Alps. Mist and cloud lie in the intervening valleys, broadened and deepened by great glaciers that have long since melted. Now, only the high slopes and valleys bear year-round ice and snow.

in its liquid form in rivers and lakes, and especially in the oceans blanketing three quarters of the earth's surface. In regions where temperatures are regularly at 0°c or below, a significant though varying amount of the earth's water is locked up in polar ice caps and the glaciers of high mountains.

Water is a nearly universal solvent and this property makes it an important force in shaping the earth's crust. Most of the minerals in the rocks and soil are water soluble to some extent. By transporting materials in solution and particles of rocks and soil in suspension from the high parts of the continents to the low, water has constantly wrought changes in the landscape through the aeons of the earth's history. The movement of water in its various states is responsible, too, for the fertility of the sea and, in large part, for the richness and productivity of the soil.

Even in terms of the earth's history, where a million years is no time at all, the changes brought about by running water are gradual, requiring hundreds of millions of years to reduce a craggy mountain peak to a comfortable, rounded hill. Far more dramatic and, comparatively speaking, rapid changes are produced by ice—water in its solid form.

Left: arch-shaped iceberg, a curious formation produced as a floating chunk of ice slowly melted. Right: lying low in the water, this iceberg off Spitsbergen appears deceptively small. Most of its bulk is in fact below sea level, and may be enough to hole a liner the size of the M.S. Europa *seen in the background. Icebergs begin as chunks of ice that break away from a glacier where it reaches the sea. Ocean currents carry some of them a long way before they melt. A few drift almost to the tropics.*

Although there have been several periods in the earth's history when ice sheets covered vast tracts of land and sea, it is the last of them, the great Pleistocene Ice Age, that left us our heritage of rugged mountain regions and polar ice caps. At the beginning of the Pleistocene, about one million years ago, the surface temperatures of the planet dropped sharply, climaxing a long period of cooling. In the ensuing millennia massive accumulations of ice spread over large areas of both temperate zones, alternately advancing and retreating, four times in all. The last retreat began at least 10,000 years ago. Because of the long time span needed for the development of glacial and interglacial periods, as the warmer periods are called, it is difficult to know whether we have at last come out of the ice age or are living in an interglacial period that has already lasted 10,000 years and may soon—say in another several thousand years—give way to another advance of the ice.

Although today's ice sheets are puny compared with the great continental ice caps and glaciers that covered much of Europe, North America, and Siberia during the Pleistocene maximums, large areas of the earth's surface are still ice-covered. The ice that now covers the polar regions and high mountain valleys behaves

Right: the Pelvoux mountain group in the French Alps. Piles of rubble are moraine materials—rock fragments torn from the mountains by a glacier that bore the pieces downhill, and dropped them where it melted. Moraines may dam mountain streams, creating long, narrow lakes that occupy glaciated mountain valleys. Part of a glacial lake takes up the foreground here. Below: this Greenland scene shows some of the processes just described in action. The immense glacier that once occupied the central valley has receded, but lesser glaciers still flow in from the tributary valleys. As they reach the central valley, these glaciers spread out as tongues of ice, blackened by the bits of rock they have transported. Rock debris lies particularly thickly around the melting edges of the glaciers and creates a sheet of moraine covering the floor of the major valley. Meltwater trickling from the glaciers forms streams that carve channels in this loose material, some of which the waters bear away. The whole landscape is still emerging from an ice cap that planed down and rounded off the summits of the mountains that rise above the glaciated valleys.

similarly to the glaciers of the Pleistocene period. Thus scientists trying to fit together a theory of past glaciation do not have to rely only on the clues etched into rocks by the ancient ice sheets and the piles of debris left behind, but they can also study the behavior of today's polar ice caps and active mountain glaciers.

The polar ice caps are vast storehouses of cold, profoundly affecting the circulation of the atmosphere and oceans and, consequently, the world's weather. In addition, these great masses of ice also lock up something like three quarters of the planet's fresh water. Thus, 10,000 years after the retreat of the last major Pleistocene ice sheets, ice still plays an important role in modeling and shaping the continents and the distribution of living organisms, including man.

During the period of maximum Pleistocene glaciations, the ice must have been static at times. When the growth of ice sheets at their centers was balanced by peripheral melting, the ice neither advanced nor retreated, but even so it was shaping the land. As the ice moved outward from its center of accumulation to its

Fertile farmland in Kansas, part of an immense expanse of flat land in North America enriched by rock debris dumped by ice sheets. Some of this so-called glacial drift still remains where the ice sheets melted; some was washed far away by streams that flowed from the ice. "Rock flour" or loess, windblown from glaciated areas, covers much of the Missouri and Mississippi valleys.

melting edge, it carved out valleys and smoothed off the tops of mountains beneath it. Its great weight depressed the continental rocks of the crust, which is still recovering. The rock floors of the Scandinavian peninsula and the sea around it subsided some hundreds of feet. Now that this area is relieved of its glacial burden, it is rising at a rate of about three inches a century. Eventually, as the crust springs slowly back, this rise will cause the Baltic Sea to flood parts of Denmark, Germany, and Poland.

Although the Northern Hemisphere's ice sheets cover only comparatively small areas, that of the Southern Hemisphere blankets an entire continent, burying all but the tallest pinnacles of its mountain ranges under thousands of feet of ice. The massive weight of the Antarctic ice depresses the continent, and sliding and grinding seaward, carves and shapes the land beneath. If the Antarctic ice sheet were suddenly to disappear, the continent that emerged would show the classic features of a landscape sculpted by ice: wide, deep valleys and round, polished mountain tops. The meltwater would raise the depth of the oceans by 200 to 300 feet, flooding large areas of the world including many of our largest cities, New York, San Francisco, London, Amsterdam, and Copenhagen to name but a few. The crustal rocks, released from their burden of ice, would move unpredictably upward causing corresponding, and possibly disastrous, shifts in the crust the world over.

The Greenland ice cap, which is very much smaller but in many parts as thick as the Antarctic ice, weighs 2500 million million tons. It has depressed a vast area of the land lying beneath it well below sea level. Freed from the ice, Greenland would begin to rise just as the Scandinavian peninsula is doing today. Continents do not, however, emerge suddenly from under ice caps. Although the ice sheets and mountain glaciers have shown signs of both increasing and decreasing, there is no sign that the Antarctic or Greenland ice caps are beginning to disappear. Should they melt, however, the emerging land would be gentle and its mountains smoothly rounded.

Most of the world's high mountains are not smoothly rounded, however, but sharp and angular. The smooth contours left after the retreat of the last Pleistocene ice sheets were soon attacked by the effect of daily and seasonal temperature changes that alternately heat and cool the rocks. Because the minerals in the rocks expand and contract at different rates, the rocks break and crumble. In the higher latitudes and on mountain slopes, frosts add their toll. At night and in winter, the sharp report of splitting rock is frequently heard on mountainsides. Some of the loosened rock falls to form conelike slopes of shattered rocks known as *talus* or *scree*. Other weather-loosened rocks are brought down in the snow avalanches and by the glaciers.

Although we do not know what causes the large-scale surface temperature fluctuations that ultimately give rise to ice ages, we do know that ice sheets and glaciers will form when the amount of snowfall exceeds that which is evaporated or melted. If this persists long enough, ice caps and glaciers will appear. At present these requirements are satisfied at various altitudes above sea level depending on the distance from

Above: telltale signs of glaciation in Waterton Lakes National Park, Alberta. The round cirque lake in the foreground occupies a hollow carved in rock by the alternate freezing and thawing of ice. The gently curved slopes beyond the lake show where a glacier has broadened and deepened a mountain river valley, changing its V-shaped cross section into the typical U-shape of a glacial valley.

Left: bedrock above a mountain stream owes its smooth, polished appearance and deeply scored parallel grooves to the abrasive action of rock fragments embedded in ice. Such glacial scratches can be seen wherever glaciers have scraped bedrock bare, from the Antarctic Peninsula to New York's Central Park.

the equator. The nearer the equator, the higher up the mountain will be the glacier, while near the poles glaciers will be found at sea level. The line above which there is perpetual snow is called the *snow line.*

The amount of snow falling above the snow line varies according to local wind conditions, sunshine, and atmospheric moisture. Any mountain wholly above the snow line will be almost completely snow-covered, whereas a mountain with just its top above the line will have only its peak dusted with snow, although there might be small accumulations in cracks and gullys.

Snow falls when the moisture in the air is cold enough to form crystals. When snowfall is heavy and prolonged the beautiful, hexagonal crystals are forced closer together and are eventually crushed by the weight of accumulating snow above them. At this stage they form a compacted snow called *névé.* The texture of névé—the first stage of ice formation—is like snow that has been squeezed into a snowball. Although ice has a crystalline structure it behaves almost as if it had no regular form at all. This quality enables it to flow, not like liquid water but rather like cold pitch.

As ice builds up on mountains and high plateaus it presses downward and outward in all directions into the valleys beneath. In both Greenland and Antarctica it continues to flow until it reaches the edge of the sea where stresses set up by tides and ocean currents break off large chunks to form icebergs.

The Antarctic ice sheets flowing into the sea may have fronts, or shelves, hundreds of miles long reaching 100 feet or more above sea level. The icebergs breaking off these shelves may be from 10 to 20 miles long and from five to 10 miles wide. Because they extend so far below the surface, they are often stranded on the sea floor. As the ice melts, however, they float and drift great distances on the ocean currents before they melt completely.

Provided the conditions for snow accumulation are met, glaciers are formed wherever mountains extend above the snow line. The higher the mountains extend above the snow line, the greater the glaciers. Mountain glaciers, found on all continents except Australia, may be well over 1000 feet thick and range up to 75 miles in length and several miles in width. The largest are found in southern Alaska where some peaks extend 15,000 feet above the snow line. There are spectacular alpine glaciers in Switzerland and France, but the best-known of all glaciers is that flowing from the Western Cwm on Mount Everest.

Glaciers on steep slopes move faster than those on shallow slopes and, because of friction generated between the glacier and the bedrock and valley walls, the surface moves faster than the foot and sides. As a glacier reaches lower and warmer altitudes, it loses some of its ice through melting and evaporation, but will continue to advance as long as the ice moves faster than it melts. Many glaciers extend for miles below the snow line before reaching a level where the ice melts as quickly as it moves.

As a glacier moves down its valley, weight, stresses, and friction may cause cracks, or *crevasses,* several feet wide and maybe hundreds of feet deep to appear suddenly and then to close just as suddenly. This makes the entire glacier surface dangerous for mountaineers. A glacier flowing over a sharp change of slope in its valley floor will form an icefall rather like a frozen waterfall, riddled with hundreds of crevasses and covered with towering ice blocks that break off and thunder down to the lower level.

These rivers of ice, cutting and grinding their way down mountain valleys, remove loose bedrock and push it, and any other stony materials in their path, ahead of them. When these materials of all sizes are released at the ice front, they form huge piles called *moraines.* When excavations uncover such debris, geologists know that they are looking at the remains of a glacier of long ago. The rock pile left behind after a glacier's snout has retreated is called a *terminal moraine,* and that formed by a glacier pushing the rubble to its sides is a *lateral* or *side moraine.* A third type, *medial moraine,* is found where two glaciers came together or at the point at which a single glacier flowed around an obstruction.

When glaciers retreat up their valleys, their moraines are often wholly or partially swept away by the rivers of meltwater flowing down the glacial valley. Eventually the moraine material is distributed along the length of the valleys and over the lowlands. A continental glacier may leave behind a mass of finely ground rock—*loess* or *rock flour*—covering hundreds of square miles.

The pyramidal Matterhorn viewed from Zermatt in Switzerland. Freezing and thawing of snow on its slopes shattered rocks, carving ice-filled hollows that dug deeply into its flanks, which became steep rock walls separated by frost-sharpened ridges.

Flash flood in Arizona's Monument Valley. A distant cloudburst may have sent this torrent of water swirling through a normally dry riverbed in the desert. Before the waters dry up they will have removed rock particles from the sides and the floor of the channel, borne them downstream, then dumped them. In such ways flash floods and perennial rivers eventually carve out valleys.

Winds sometimes distribute the loess over large areas and deposit it to form such mineral-rich soils as are found in the Great Plains regions of North America, which are some of the most productive wheat-growing areas of the world.

The exposed rocks of U-shaped glacial valleys appear smooth, but a close look at them may reveal many parallel scratch marks, or *striae*, etched in them by rocks and boulders carried over the bedrock by the moving ice. The boulders and rocks that were ground against the valley floors have a flat almost smooth surface on one side in contrast to river or marine boulders and pebbles, which are nearly always smoothly rounded all over.

Higher up the valleys, where frost action and glacial erosion tend to produce very steep rock walls, deep basins or *cirques* are formed. Two cirques on opposite sides of a peak define a sharp narrow ridge called an *arete*. The most spectacular combination is a needle-pointed pyramid produced by three cirques cutting into the same mountain peak. The Matterhorn, near Zermatt in Switzerland, is thought to be the most perfect of all such peaks. When the glaciers disappear, rainwater and melting snow fills the old cirques to produce steep-banked lakes, or *tarns*.

Small streams overflowing from these lakes run into the small, high valleys once filled by tributary glaciers and thence down into the valleys where the main glaciers once moved. During heavy rainfall, temporary mountain streams may swell the established streams, but at other times streams can dry up and leave the bare rock of their beds exposed to further attack by the next rains and also by wind and frost, and possibly by living organisms. Running water is also an effective landscaping agent, for ultimately it will level mountains and deposit their materials in the valleys and shallow seas. Rain not only washes loose materials off the surface, but it also attacks the bedrock itself. Rainwater that has become acidic by dissolving carbon dioxide or sulfur dioxide as it falls through the atmosphere may dissolve the soluble components of rocks, and swiftly running water abrades the bedrock with the materials it carries.

Mountain streams and young, fast-flowing rivers excavate deep, steep-sided beds. The banks, always exposed to flood erosion, are also attacked by rain, frost, wind, and living organisms. The combined work of these agents produces the typical V-shaped valley of a youngish river. The burden of rock and silt eroded from such a river's sides is carried away by its flowing waters and acts as an abrasive to grind further into the bedrock. Particularly heavy rains occasionally result in flash floods that can move mud and even huge boulders along the water courses, frequently causing severe damage to villages along the stream's bank. Over thousands of years, the river digs its bed deeper. As it does so the steep, newly formed sides continue to collapse. The digging flattens the slope of the river's course and the never-ending bank erosion broadens it until, after some millions of years, it flows slowly through a wide valley, or floodplain, of almost imperceptible slope.

The early stages of a river's development are dramatic, and in some regions we often find

Aerial view of Horseshoe Falls, the spectacular Canadian section of Niagara Falls. The falls recede as water cascading over the hard limestone lip erodes soft underlying shales and sandstones. The lip may have retreated a distance seven times the height of the falls since their discovery in 1678. The river erodes most vigorously in the center, thus accentuating the horseshoe shape.

spectacular scenery, especially when the level of the valley breaks sharply to form waterfalls. Tons of water cascading over the brink of a waterfall and charged with stones, sand, and other silts can cut away the bedrock beneath. Although it depends on the hardness of the bedrock, there are very often holes cut into the river bed at the bottom of a waterfall as deep, or almost as deep, as a waterfall is high. Waterfalls constantly cut back through the rock of the riverbed, particularly when the hard rock of the bed is underlain by soft rocks that have been exposed by the action of the fall itself. In such cases water with its load of abrasive materials cuts into the softer rock and undermines the hard rock bed. The lip of hard rock so formed will collapse under its own weight and that of the water. Bit by bit, waterfalls move farther upstream. One of the most spectacular examples of this kind of waterfall is the Niagara Falls between Canada and the United States. Niagara, which has been falling roughly 10,000 years since the retreat of the last glaciation, has moved seven miles upstream and is still cutting its way up the Niagara River at a rate of about one foot a year.

Above: satellite view of the Nile Delta—the lush, green, roughly triangular region. Here, silt dropped by Egypt's Nile River has built a great, fertile oasis jutting into the sea from an otherwise arid land. Deltas grow where silt-laden rivers flow into a shallow sea or lake that is free from the strong currents that stop silt collecting.

Left: sunlight glints on the snaky course of the Uruguay, a big South American river crossing its floodplain on the way to the ocean. Here, almost at sea level, the slow-flowing river no longer deepens its bed by downcutting, but its current eats into the outer bank of each bend so that slight kinks have evolved into huge loops.

As a river moves downhill to the sea, it picks up and drops its sediment load again and again. As a river's course flattens, its speed drops and so does the size of the rocks and particles carried. The heavier boulders are deposited first, followed by large stones, then pebbles, and so on. At the end of its run the now sluggish river deposits the smallest and lightest particles, sand, and finally clay. Of course, a fall in its volume or speed of flow may cause a river to lose its heavier particles farther upstream. Sometimes, because of an obstruction or change in direction, areas of slow water may develop in which sediments settle out in a pattern graduating from large stones to small pebbles and ultimately sand as the water speed slackens. Similar patterns may also be found where the river has flooded over its banks. This gradual sorting is disrupted if the floodwaters run swiftly, and we find larger stones deposited over smaller ones. When the river reverts to its normal flow, sand covers all the stones.

Old rivers, like old men, meander and the course of the river will wander across its plain in a series of loops. Some *meanders*, as these loops are called, are so exaggerated that the river virtually doubles back on itself. At such points, it may break through the narrow end of a loop to its lower reaches, leaving behind what are called *oxbow* lakes because they resemble the yoke that was used to hitch oxen to the plow. Some oxbows may take hundreds of years to form, but others develop within a few years. Eventually they silt up, leaving shallow depressions in the land.

Rivers carrying heavy loads of sediment into shallow coastal waters frequently build up new land at their mouths in the form of triangle-shaped *deltas* (named for the Greek letter Δ). Here again land-building is a function of stream flow. As the rate of flow slackens near the river's mouth, the water drops its sediment load. If wave action, tides, and current permit, a barrier is formed in the old channel, and the river, swerving seaward around it, cuts a new outlet. If this process is repeated often enough the typical fan-shaped river delta will form. At the mouths of the Mississippi, Nile, and Rhine, to name a few rivers where barrier formation has been repeated over and over again for thousands of years, channel-riven deltas cover hundreds of square miles of old sea bottom.

Delta soil, a rich deposit of mineral nutrients and organic debris washed out of upstream soils, may support both a thriving agriculture and a prolific fishery. Young rivers wear down the high ground and old rivers lay down new ground and distribute it over their flood-plains and in their estuaries. This action has made and still makes the floodplains and estuarine areas of the world the most important for man. Our civilizations were born around, and for the most part are still dependent upon, the great rivers.

Although most rainwater runs off into the rivers or evaporates directly into the atmosphere some is absorbed into the soil. Here it either enters the living world through plants and animals or is absorbed by rocks beneath the soil, depending on their *porosity,* or water-absorbing capacity. The degree of porosity—actually a measure of the volume of rock occupied by pore space—varies from as little as one per cent in granites to as much as 30 per cent in sedimentary rocks such as sandstone. Porous materials are generally pervious, that is, water will sink into or through them. But although clay and some other materials are porous, the individual pores are so small that it is difficult for water to pass through and therefore they are impervious. In contrast, some non-porous rocks are fissured and cracked, allowing water to pass through. Rain falling on sandstone or other pervious rock works its way downward through the pores until it reaches a layer of impervious rock, such as granite. Water then begins to fill the pores at the bottom of the pervious layer until, in the course of several rainy seasons or an unusually rainy year, the rock may become saturated. This rock-held water is known as *groundwater.* Its upper level is the *water table.*

The depth and extent of the groundwater depends on the thickness and slope of the porous rock, the amount of rainfall, the type of vegetation covering the soil above, and the amount of water already in the ground. If a porous formation is sandwiched between two impervious layers, water entering it at some point where the porous rock is exposed will seep through the entire layer to form an *aquifer.* In some parts of the world aquifers extend underground for thousands of square miles.

One extensive aquifer is the more than 100-foot-thick layer of sandstone that outcrops at the eastern base of the Rocky Mountains and slopes eastward beneath the plains of North and South Dakota, Nebraska, and Kansas. A hole drilled into an aquifer such as this usually produces a flowing, or *artesian,* well. The reason is that water in the aquifer—called an *artesian formation*—is under pressure caused by the difference in height between the altitude of the outcropping where the water enters the aquifer and the location of the borehole. When the pressure is released by drilling into the aquifer, the entrapped water is pushed to the surface by the pressure of water behind it. Water in wells sunk into an aquifer running at the same altitude over its entire extent does not flow, but rises only to the level of the water table around it and may have to be pumped to the surface.

Many limestone areas are famous for their caves. Although limestone is not a porous rock it is split by both deep vertical fissures and horizontal cracks between its layers. Because water always contains carbonic acid—formed as rain dissolves carbon dioxide in the atmosphere—it slowly enlarges the fissures by solvent action. In time they may grow into large surface

A fantastic array of stalactites hangs down from the roof of a French limestone cave. Stalagmites rise from the floor. Both consist of calcium carbonate deposited by dripping water that contains substances dissolved from the limestone rock above.

openings leading to underground cave systems extending for miles. Although the most renowned cave systems, such as the Mammoth Caves in Kentucky, the Carlsbad Caverns in New Mexico, and the Gouffre de la Pierre Saint-Martin in the French Pyrenees, are limestone formations, caves may also be formed by the action of groundwater on other soluble rock formations, such as gypsum or common salt.

Caves often contain layers of debris brought in from outside and they may be rich storehouses of fossils. Much of this material was carried in by the flowing water that formed the cave in the first place. In some cases plants and animals and even humans were swept into caves or trapped there during floods. At times, particularly during the warm interglacial periods of the Pleistocene, both animals and humans used dry caves as dens or homes. Some Pleistocene animals, such as the cave bear and cave lion, have been given the prefix "cave" to denote that their remains have been found in caves. The so-called cavemen left behind their skeletons and artifacts, and their drawings such as those on the walls of the famous limestone caves near Altimira in Spain show us that these men had developed art forms that

Hot, subterranean water produces some striking effects on the earth's surface. Above: tabular mineral deposits left by water vaporizing from a geyser in Yellowstone National Park. Right: vapor plumes in New Zealand's North Island mark boreholes that release underground steam, used to generate electricity.

made sophisticated use of form and color.

In some parts of the world caves are still used as homes or, more often, as storage places for food and wine. Subterranean rock rooms are especially suitable for wine storage, because at a depth of about 50 feet their temperatures remain close to the average yearly temperature of their regions. Groundwater, too, remains at this temperature, which explains why the water from springs is usually cool and refreshing in the summer and rarely freezes in winter.

Below 50 feet, however, heat from the earth's interior begins to raise the temperature of the rocks and warms the groundwater. Well water or spring water from very deep aquifers may be much warmer than either the surface water in the region or the groundwater closer to the surface. This earthwarmed water is the source of the hot springs that occur in many regions of volcanic activity. Hot springs often occur because hot volcanic rocks lie close to the surface. In places they may be hot enough to boil the water as it rises to the surface. Boiling springs and geysers occur when deep groundwater heated to boiling point by hot rock begins to expand. If its expansion is limited by the pressure of the water above, it becomes *superheated* (that is, reaches a temperature greater than its normal boiling point). As its temperature continues to rise, the superheated water expands and pushes the cooler water above it to the surface. With the pressure above relieved, the superheated water explodes as a column of steam, blowing out the remains of the water column above it as a geyser.

Before water can begin its work of smoothing the earth's crust, it must be lifted to high ground where it starts its downhill flow. This, as we have seen, is achieved by a combination of the heat of the sun and the movement of winds in the earth's gaseous envelope. Wind, however, plays more than a passive role in the story of the rocks.

In many arid regions wind so efficiently removes loose materials formed by weathering that only deserts of boulders and bedrock remain. In most of the world's dry areas, however, wind removes only the fine silts, and in such areas we find sand deserts. During periods of prolonged drought, however, winds have been known to remove practically all the topsoil from semiarid regions that have been denuded of grass by damaging farming or grazing practices, and carry it for thousands of miles. This

happened in the Great Plains of the United States during the 1930s where dust storms caused such fantastic losses of soil that an area of many thousands of square miles became known as the Dust Bowl.

Loaded with sand and dust, winds are powerful abrasive agents. The principle of wind erosion is similar to sandblasting, in which a powerful stream of air and sand blasts grease, rust, or soot off a surface needing cleaning, taking some of the base material along with the dirt. Wherever rock is exposed to windblown sand, the effects of natural sandblasting are particularly noticeable near ground level, because sands are heavy and not lifted very high by wind. Wind erosion on soft rock has carved the bizarre and beautiful formations to be found in the American Southwest, northern Mexico, the Middle East, and North Africa.

In the loose-sand regions of the desert areas of Africa, Australia, and the Americas, persistent strong winds form sand dunes. Whether in a desert or on beaches, a sand dune usually starts when wind-blown sand piles up against an obstruction such as a boulder, a piece of driftwood, or even a plant. Although some dunes may be only a few feet high, those of the Sahara's loose-sand regions often rise to more than 1000 feet and extend for several miles. It is estimated that dunes now cover some 300,000 square miles of the

Left: wind takes a hand in shaping Monument Valley's weird desert landscape. First, wind removes particles loosened from rock by expansion and shrinkage caused by heating and cooling. Armed with such particles, the wind sandblasts rocks near ground level. As they rub on rock and each other, the windborne particles lose their sharp edges and become rounded sand grains.

Below: moving air molds and remolds the land surface in White Sands National Monument, New Mexico, where wind piles loose gypsum particles into changing patterns of sand dunes. Dune formation may be influenced by quite trivial obstacles. Here, grasses trap sand blowing from right to left, creating marked depressions in the level of sand in their lee.

Sahara and are known to be increasing every year as the great desert spreads southward.

In the Sahara and other loose-sand deserts dunes often migrate downwind 100 feet in a year. The action is exactly like that of drifting snow. Wind blowing against the windward side of a sand dune carries loose sand over the crest and drops it on the leeward side. Sand dunes moving this way have been known to bury forests, farms, and even towns. In some threatened areas their migration has been halted by planting resistant grasses and shrubs with roots that hold the surface materials together.

However, nowhere is the dramatic earth-leveling interplay between the water world—the *hydrosphere*—and the atmosphere staged more grandly than on the seashores. Traditionally we think of the coasts as the battleground between land and sea, and certainly it is here that we can observe some of the most destructive effects of water on the crustal rocks. Coasts, however, can also be areas where new land is built up by the agency of ocean currents. It is in the coastal waters that we find the greatest variety of marine life, and between the marks of high and low water, the *intertidal zone*, we can see some of the signs of life's struggle to survive, for organisms in this no-man's-land between sea and continent live part of their lives in air and part in water, and have had to adapt to a wide range of environmental conditions, pressures, and temperatures.

Any coast is besieged by an alliance of air and water in an endless struggle with the shore. The course of the events that occur and their effect on the shape of the shore depends upon the hardness of the particular rock involved, on the nature of

Waves break on the coast of volcanic Lanzarote, one of the Canary Islands off northwest Africa. Bits of rock picked up by the waves batter the shoreline, chipping off further pieces. In such ways the destructive force of the sea acts to wear away land.

Above: sand spit at the mouth of the Minnamurra River, southeast Australia. Sediment deposited in the calm waters behind a rocky, eroded foreland has been piled up by sea and river as a sandy tongue above sea level. While the sea devours part of a coast, it may thus help to build land outward elsewhere.

Right: dune-killed trees near Western Brook, Gros Morne National Park, Newfoundland. Coastal dunes grow as onshore winds blow dry beach sand inland. Sometimes the sand may form dunes several hundred feet high. As it migrates inland, a dune may bury trees in its path, killing them before it has passed on.

the sands, and on the ability of both to withstand the combined pressure of wind and water.

Waves, the most obvious of the coast-shaping forces, are born in the oceans through the interplay of wind and water. Strong winds, blowing over vast stretches of ocean, produce regular, rhythmic movements in the water. The stronger the wind the greater the rise and fall of the water, and it is this rise and fall that we recognize as waves. Most waves are small affairs, but during storms they regularly reach 60 feet or more in height and can, on occasion, touch such dizzy heights as 100 feet and more. It is not merely the height of the wave that is important but also the wavelength—that is, the distance between wave crest and wave crest, which in storm waves may be more than 500 feet. A third factor is the speed of the wave passing through the water, and during storms this is frequently around 60 miles per hour. When waves hit a coast, they usually break and roll up the beach with a thundering roar, or if they are in deeper water break directly on the coastal rocks.

The reason why waves break in shallow water is simply that the bottom of the wave, scraping on the shelving shore, is slowed down, while the upper part hurtles along at full speed. The wave literally breaks up, hurling its water onto the shore. As the breaking wave scrapes the bottom, it pulls sediments up with it and throws them ahead as it breaks. To this action we owe the formation of sandbars and lagoons. If the pattern is repeated often enough, the sandbar may become an established part of the shoreline or may enclose a lagoon between the shore and the bar. One of the most famous examples is Fire Island and the Hampton Beaches off the Atlantic coast of New York. Lagoons may in time become salt marshes, and after a long period of in-filling may eventually dry out to form part of the land itself.

If currents and waves continue to pile sediments along the shore, the prevailing winds may carry them farther inland in the form of sand dunes. In this build-up, sediments that may originally have come from mountains far inland to find their way to the sea by way of the rivers and estuaries return to the land. Here again we have an example of land building under circumstances in which the sea seems to counter the erosive action of the rivers and streams. Changes in wave and wind patterns cause corresponding changes in the movements of the sediments. Their build-up in a particular area can quite easily be reversed and they can be removed, or deposition may begin in an area where the coast was previously being worn away.

On rocky coasts the sea may change its action

A natural arch wave-worn in chalk cliffs at Étretat, northern France. Such arches originate as caves gnawed from weak layers of rock, especially layers enfeebled by fissures. If two such caves work backward—one from each side of a headland—an arch is formed where they meet. As erosion continues, the roof of the arch collapses, leaving the seaward side as an islet or "stack."

much as it does on the shallower sandy coasts, but because rocky headlands and coasts generally stand in deeper water, they are subject to constant erosion by waves. The very power of the pounding waves—tons of water at high speed—may shatter or loosen the rocks to produce the massively carved faces of the sea cliffs. Just as in the rivers, the wave-tumbled rock fragments will serve as tools with which the waves can cut and undercut the cliffs from which the rocks originally came. When the cliff is sufficiently weakened it falls into the sea, starting the whole process again. Sometimes the sea dissolves away the softer rocks, leaving the harder rocks standing for a while, providing us with the familiar sea-caves, arches, and stacks.

At the very end of Europe, at the most southerly point of Portugal, stands the famous Cape St. Vincent, familiar to ships plying between North America, northern Europe, the Mediterranean, and Africa. Here, where the mighty Atlantic meets the resistance of old Europe, the sea has carved a series of capes, magnificent cliffs, rock stacks, and caves. One of the most spectacular of the latter is the *forno* on the Sagres peninsula. Waves rolling into a cleft beneath the peninsula force the air out of a vent that has been formed on the top of the cliff, producing the most eerie of sounds. In fact, along the southern coasts of Portugal can be seen almost all the features that can be expected in coastal zones, from arches and stacks to sandbars and lagoons.

So once again we see the play between water, air, and land: the effect of pressure, the unique quality of water both as a solvent and as a carrier of materials to either erode the land or build it up anew.

If our planet were built on a logical plan we might expect that the lighter gaseous elements would be farthest away from the center and that water, the next most dense, would form the second layer, and so on in order of increasing density to the core. A planet so organized would have its crust submerged beneath the less dense water of the sea. And in fact, for over 70 per cent of the earth's surface this expectation holds. Yet, for some reason, and in some places, the earth's crust is thrust above the ocean's surface. As we have seen, whenever this occurs there is a constant grinding down almost as if there were some intelligence or law—shall we call it nature—that says that the earth's crust has no place except under the ocean.

A coral "mushroom" near Mauritius in the Indian Ocean. Marine erosion has undermined an old reef, to leave only this top-heavy, scrub-covered relic poking up above the level of the sea.

This householder bleakly surveys his garden, which collapsed as the sea undermined a nearby cliff on England's east coast. Soon his house, too, will fall victim to the encroaching sea.

Pluto's Workshop

If weathering and erosion were to continue without interruption by counteracting forces, the earth would be a very different place, for during its long history there has been more than enough time for these processes to have worn down the continents many times over. In fact, the geological record suggests this has nearly happened once or twice, but the powerful forces deep within the earth that are part of the planet's inheritance have always frustrated the agents of external destruction.

Before we look at the cycles of energy buildup and release within the earth's interior and the pressures and crustal movements they cause, we ought to know something of the fate of the sediments stripped off the continents by erosion and deposited in the seas by the action of the wind and water. Only the finest sediments travel far enough to be deposited in the seas of the continental shelves; few are found in the really deep ocean basins and then only in limited regions. As a result there are thousands of square miles of deep ocean bed that have little or no continental sediments lying on them. Instead, their sediments are composed of the shells and skeletons of marine organisms that rain down from the waters above. This is not to say that the oceans contain nothing of continental origin, for their very saltiness is derived from minerals of the continental masses.

In the shallow seas bordering the continents, the bedrocks of the sea floor dip to form basins as if to receive and store the sediments eroded from the continents. It is not known whether these depressions in the crust, called *geosynclines,* were formed before the sediments ran into them or are produced by the weight of accumulated sediments deposited in them. But leaving aside the mechanism of their formation, many are known to be very deep, and to extend over hundreds of square miles, and to be capable of holding layers of sediments thousands of feet deep. The in-filling of

Artist's impression of the turmoil on the earth's surface while land, sea, and atmosphere were in their infancy. Lightning played and the seas steamed as frequent volcanic eruptions threw new, molten rock onto the still unstable crustal surface, and belched dark clouds of ash high into the atmosphere.

D.Nockels

a geosyncline with sediments takes millions of years, during which the general global—or regional—climate may change several times. Such climatic changes are reflected in the nature of the sediment deposits, a fact that permits us to reconstruct at least segments of the earth's climatic history. The sediments laid down in periods of extremely arid conditions are found to be composed of sharp-edged and angular particles, indicating they were derived from desert areas and deposited by the wind. Sediments deposited during other periods, when the land was covered with luxuriant vegetation and abounding in animal life, contain the remains of plants and animals mixed with rounded, water-worn particles deposited by rivers and streams.

In time, the weight of the accumulating sediment pile pressing on the lower layers, acting together with the heat from the earth's interior, cements the particles into new rocks, the *sedimentary* rocks. Because the original sediments were deposited in layers, these new rocks are also layered, or *stratified*. In some cases, particularly in volcanic regions, heat and the intrusion of molten rocks into the sedimentary rocks melts them and changes them into rocks of a different nature, the *metamorphic* (or changed) rocks.

After several million years of crustal subsidence, the geosynclines and their layers of sedimentary rock are pushed up again. Uplifting is a slow process but more rapid than the action of the erosive forces that attack new rocks when they rise above the surface of the sea or the land. As soon as they form slopes that provide the means for water to act under gravity, the loose, uncompacted sediments on top are washed off. The cycle then repeats itself as weather and water begin their work of breaking up the new sedimentary layers. Even though they are harder, any exposed metamorphic rocks will also be subjected to weathering by frost, ice, and snow and to the abrasion of wind and water. And as the land continues to rise, new highlands and mountain ranges are born.

It is strange that these countering forces should exist. Perhaps they are interlinked in a major cycle of crustal movements the nature of

Stripes on these steep cliffs are layered sedimentary rocks laid down over many millions of years and bared by the powerful erosive force of the Colorado River. Rock layers totaling 95 miles thick in some places have been formed since the earth began.

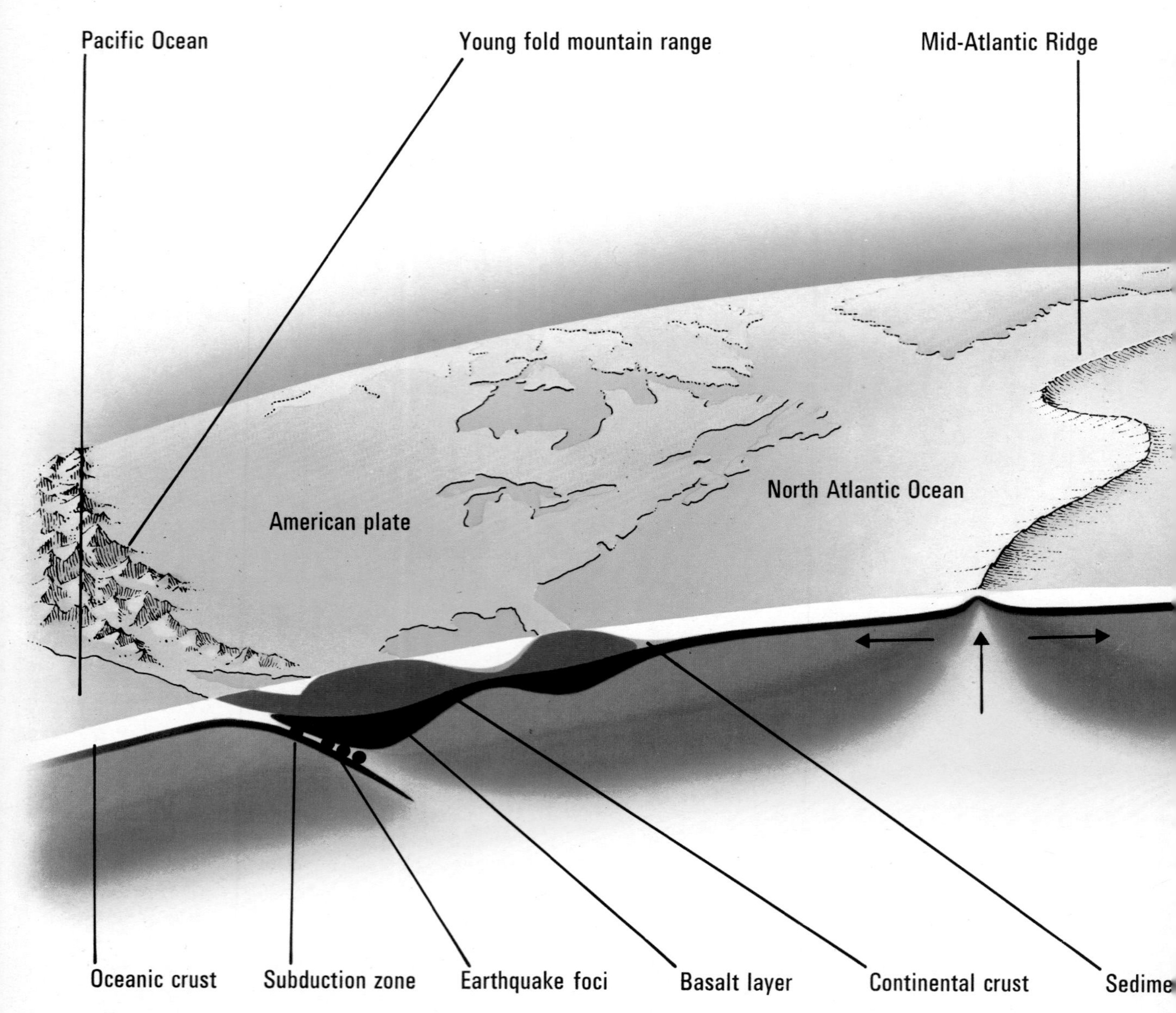

Section through the earth's crust (vertical scale exaggerated) shows forces that shift continents, build mountains, and trigger earthquakes. The relatively light granite and basalt of the continental crust floats on the denser mantle that lies beneath the so-called Mohorovicic discontinuity. Mantle material upwelling in the Atlantic thrusts up the Mid-Atlantic Ridge and forces continental plates apart. The American plate overriding the Pacific Ocean plate at a so-called subduction zone has created young fold mountains and earthquake zones. Collapsed crust between Africa and Arabia forms the Red Sea, a part of the Rift Valley.

which will become apparent only when more is known about the origin of our planet and the events within its interior. We do know that the earth's crust floats on an inner layer of much denser material, the *mantle*. Although it consists of apparently solid rock, the mantle behaves as if it were fluid, buoying up the lighter crustal rocks embedded in it much as water supports an iceberg. On the basement rocks of the crust ride the even lighter rocks of the continents.

The bulk of the continental rock is concentrated on the side of the earth away from the Pacific Ocean as if the earth could be divided into a land hemisphere and a water hemisphere that comprised the vast Pacific basin. Such a situation ought to produce an eccentric wobble in the planet's spin sufficient to cause it to disintegrate. The reason it has not seems to be that the 20-mile layer of crustal rock under the continents balances the three miles of sea, the four miles of

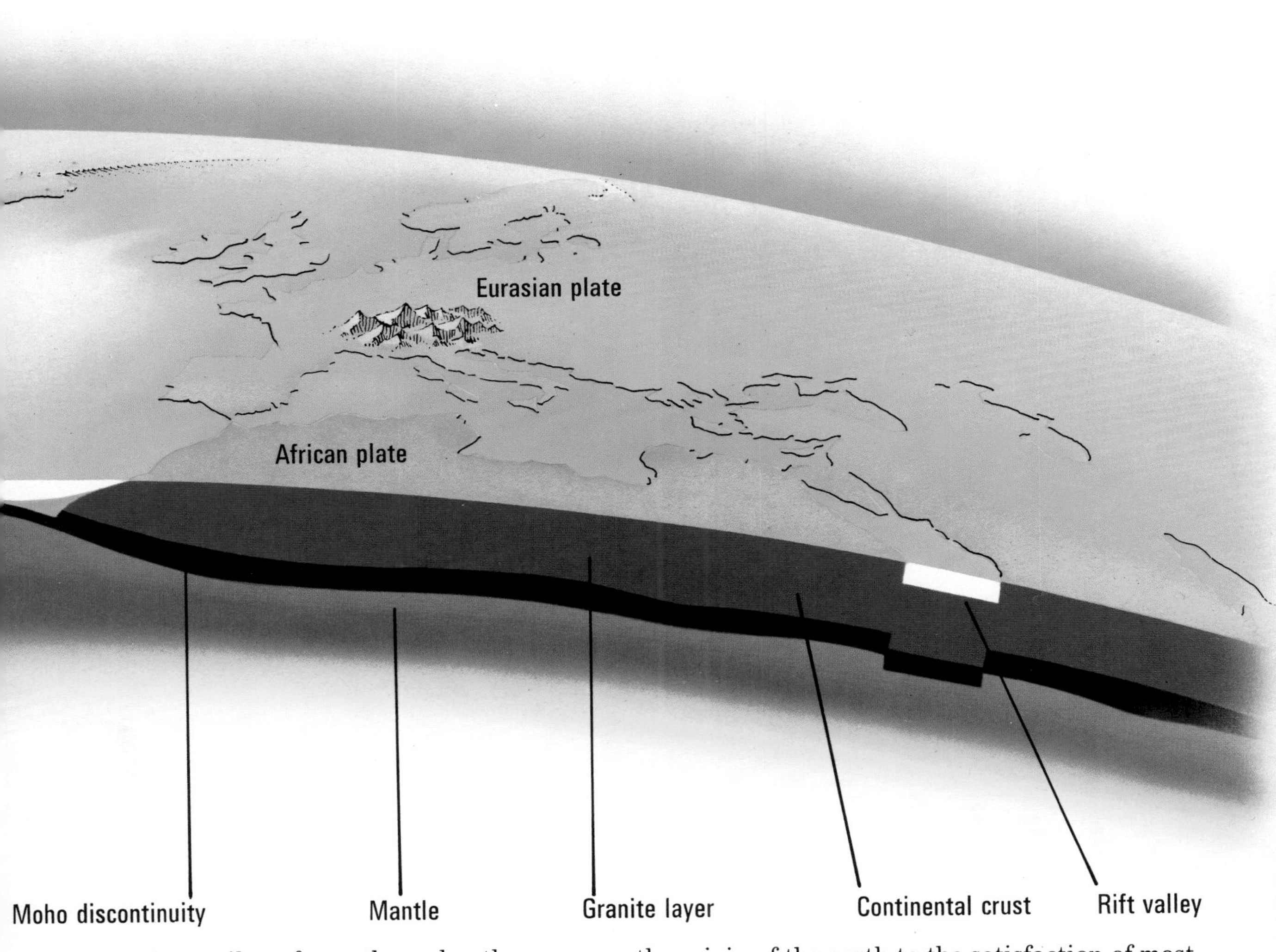

crust and 13 miles of mantle under the ocean. If we were to drill out a random series of 20-mile-deep cores from across the earth's surface, we should find that they all weighed approximately the same.

Why there should be a crust at all is a mystery and its existence poses a number of problems to scientists studying the earth's origin and composition. The crust must have formed early in the planet's development, and to understand how it came to be, we have to go back to the formation of the sun and its planetary family. From the time that man first developed the capacity to wonder, he has sought a likely explanation of the origins of his home. One of the oldest descriptions is in the Book of Genesis, which, though not explaining the origin of the earth to the satisfaction of most scientists today, seems at least to describe the sequence of events correctly.

The primitive solar system was born some 6000 million years ago when gases and dust in our part of the galaxy started to condense into a spherical cloud that began to shrink, becoming denser at its center. As it rotated around the heavier center, the cloud changed gradually into a flat disk, more or less the dimensions of the present solar system. Nuclear reactions in the dense center eventually caused it to glow, and thus a primitive sun began to give off its energy, while around the glowing proto-sun the materials in the cloud condensed into the planets.

The earth, then, along with the other planets

was formed from galactic gas and dust. As this little blob of universal substance grew larger by continued accretion, its density increased. Within its interior nuclear reactions began to generate intense heat that melted the proto-planet's component materials. The heavier ones settled toward the center and the lighter, accelerated by the centrifugal forces set up by the new planet's rotation, rose to the surface.

At that time there was no atmosphere—at least not as we understand it—enveloping the heaving, bubbling surface of the earth. As time went on the new planet lost much of its heat into space. The intensity of the nuclear reactions gradually subsided, and the surface cooled slightly.

It was during this period that the primitive crust began to appear. The lighter materials that were to become the crustal rocks bubbled to the surface in the same way that the scum or slag boils to the surface of molten iron in a blast furnace. This layer of slag—the newly forming crust—floated on the denser substances of the mantle. The lighter and volatile water vapor and gases were forced out beyond the crust but were prevented by gravity from escaping entirely. This de-gassing of the early crust and mantle gave birth to the first of the earth's atmospheres.

As the earth continued to cool, water vapor in the atmosphere condensed into liquid. But the still intense heat converted the first rain into

Volcanic activity on and near Iceland reminds us that this island straddles the Mid-Atlantic Ridge, formed where a crack in the earth's crust separates the great American and Eurasian continental plates. In 1963 magma welling up from this crack gave birth to a brand-new volcanic island called Surtsey. Left: aerial view of the island. Steam rises where torrents of molten rock reach the sea. Below: close-up view of red-hot lava sliding into the sea on a part of Surtsey's unstable coastline.

vapor again before it had fallen far enough to reach the earth's surface. Eventually, however, after the earth had orbited the sun a few million times it cooled sufficiently for rain to reach its hot, craggy surface. Still it was instantly reconverted into steam, but after millions of years more had passed, the surface cooled enough for water to remain liquid. As the rocks cooled, the first rivers ran downhill filling cavities, basins, and hollows to give birth to lakes and seas, and water then began to play its age-old part in the story of the rocks.

That early earth would have been very alien to us. The lakes and seas were hot, even boiling. Streams of molten lava similar to that flowing from volcanoes today ran between the cooler rocks, and other areas were covered with hot cindery rocks. It is likely that for millions of years the planet's surface was obscured by dense clouds of water vapor, ammonia, and methane in which lightning flashes lit the sky and thunder rolled and echoed. Beneath the cataclysmic surface massive waves, generated by the intense heat, sorted all the various materials that were present in the mantle according to their densities. If we add to all this the centrifugal pressures imposed on the planet by its rotation and orbital swing and the enormous gravitational attraction of the new sun, then in our imagination we can conceive a picture of the surface of the earth in a

The lines on this world map represent the edges of the larger of the 19 crustal plates that fit together rather like pieces in a jigsaw puzzle to make up the solid surface of the earth. Mantle materials thrusting up between certain plates force them apart and keep the whole puzzle in slow but constant motion. This shifts the continents around and causes some plates to override others. Where ocean floor is forced far down into the mantle, deep ocean trenches, island arcs, and mountain ranges are formed.

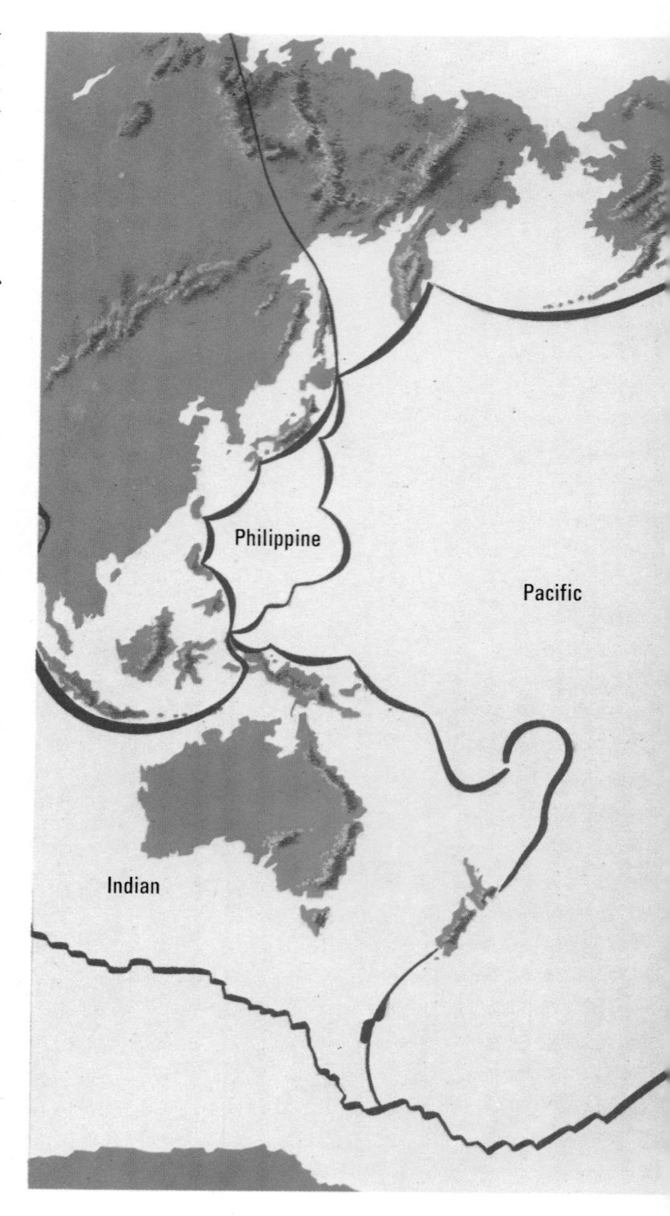

state of incredible violence for thousands of millennia.

Meanwhile rocks had begun to form. The lighter granite rocks of the continents had risen to overlay the dense basalt rock of the crust. Already the first rains were wearing these away. Loaded with solvents dissolved out of the primitive atmosphere, the rain was many times more corrosive than today. So weathering and erosion began as soon as the earth was cool enough to enable water to run down the rocks.

In the seas laden with salts that had run off the land and rich with chemicals dissolved from the atmosphere, new substances were being synthesized that, after many false starts, would lead along the road of evolution to life. The earth, torn by internal and external forces, existed in this way for millions of years. In time it settled down and assumed a structure much like that it has today: a dense molten center pulsating with nuclear energy; a mantle of solid, yet fluid, rocks wrapped around the hot core; and a thin crust supporting the lighter continental rocks.

The lighter continental rocks are not, however, uniformly distributed. One possible explanation is that early in its history the earth threw off a large portion of its lighter, outer covering. As it spun away this piece broke into two. The larger and heavier part escaped from earth's gravitational field to be pulled by the sun's stronger gravity into an orbit of its own. This daughter of the earth became the planet Mars. The other portion remained in earth orbit to become our familiar moon. The parting of Mars and the moon from the earth left an asymmetrical world with a relatively small landmass surrounded by ocean. Samples of moon rocks, which have been found to be similar to those in the earth's crust, are beginning to support this hypothesis.

Unlike Jules Verne's Professor Lidenbrock who traveled to the earth's center to observe it at first hand, we are limited to studying its structure by analyzing the information provided by natural and artificially generated earthquake shocks. Nevertheless, a great deal has been learned about earth's interior. Beneath the rocks of the crust is the dense but fluid mantle, and at the center is a molten iron and nickel core under a pressure of $3\frac{1}{2}$ million atmospheres.

To reconstruct the story of the continents we must leave the crust for a while and immediately concern ourselves with the huge convection currents flowing through the mantle; these *tectonic forces*, as they are called, are responsible for the movements of the continental masses. If we picture the crust as made up of large plates floating on the outer mantle, they can be visualized moving along with the convection currents.

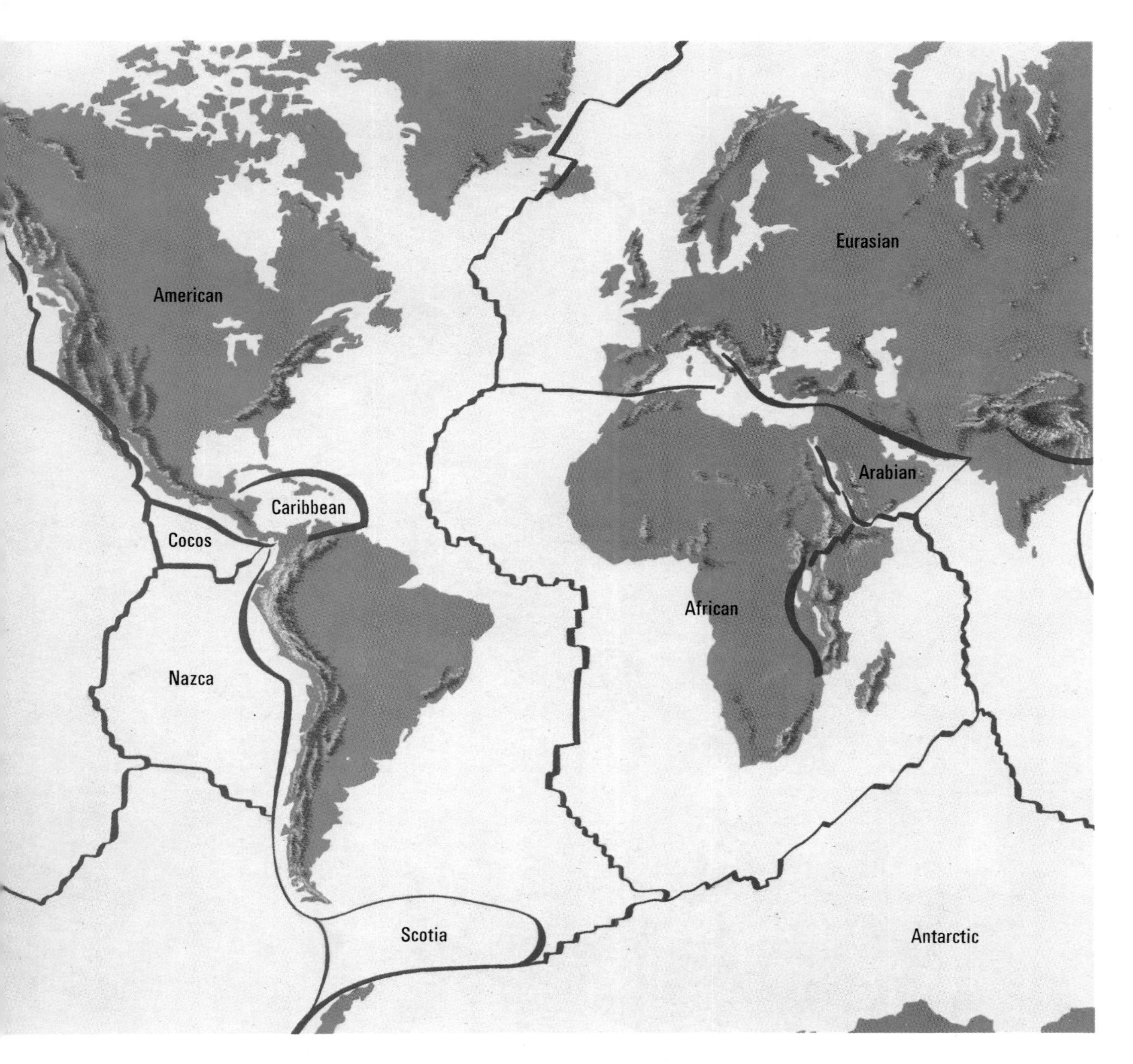

As rising currents hit the plates they are deflected, taking the plates with them. Plates moving away from one another in this way are known as *diverging plates*. Cracks appear in the crust at the sites of divergence, and *magma* (molten rock from deep within the crust) wells up through them to form ridges of intense volcanic activity. Diverging plates may move in many directions with respect to each other. For example, in the separation of North America from Europe—a process that has been going on for the last 200 million years—the two continental plates have not only moved away from each other, but have also moved northwest and northeast respectively. These movements produced the North Atlantic ocean basin. A ridge of active volcanoes in the middle of the Atlantic and lateral ridges to the sides of the main divide show the multidirectional drift. Along this diverging ridge lie many volcanic islands, such as Iceland and the Azores, whereas others, such as the Canaries, which were once located directly over the ridge, are now situated away from it, having drifted with the separating plates. The farther away from the central ridge they are, the older they are.

Intense volcanic activity in Iceland and the Azores indicates that the two continental plates are still diverging. In 1963, a completely new

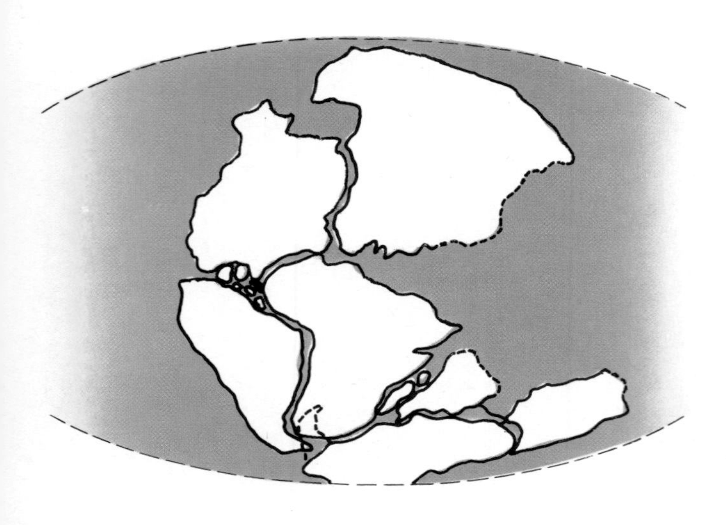
200 million years ago

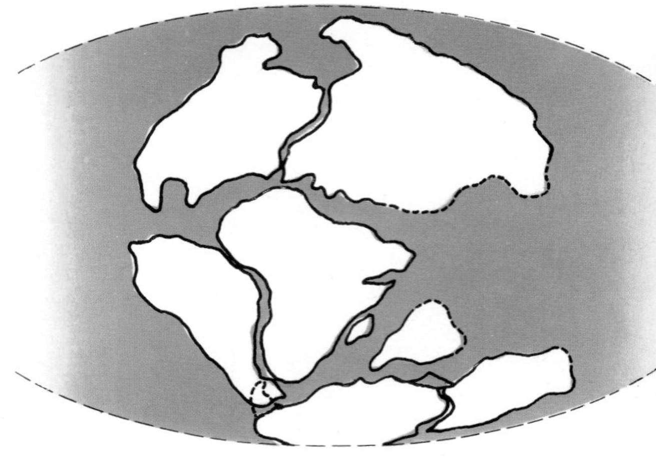
135 million years ago

Four maps suggest how continental drift has probably altered the face of the earth in the last 230 million years. Some 200 million years ago landmasses interlocked as the supercontinent Pangaea. Then 180 million years ago Pangaea began to split into (northern) Laurasia and (southern) Gondwanaland. These also broke up. By 65 million years ago the continents were assuming more familiar shapes and positions. Past interlocking probably occurred at the edges of what are now submerged shelves fringing the continents.

volcanic island, Surtsey, was formed some 11 miles off the southwest coast of Iceland as the result of undersea volcanic activity along the North Atlantic ridge.

On the morning of 14 November, 1963, red-hot lava oozed from a crack in the crust and nosed up through the layers of sediment into the ocean itself. The molten rock boiled the icy northern waters, and a cloud of steam rose up over 20,000 feet into the air. The first people to see the eruption were fishermen who at first thought that a ship was on fire, but soon realized they were watching a violent volcanic eruption. All through the first day the eruption grew in intensity and the continued flow of the lava through the fissure built a cone that soon broke the sea's surface. Explosive bursts threw great fountains of inky-black ash 1500 feet into the air. Steam carried the fine particles high into the atmosphere while the heavier ones fell back into the water. Massive bombs of solidifying lava crashed down the cone's sides or plunged into the sea. Then lava fountains began and the molten rock poured down the sides of the cone and out into the Atlantic, building a hard protective crust where the cold ocean waters solidified the rock. After 10 days the new island, now over 600 feet high and nearly half a mile across at its widest part, looked like an elongated horseshoe. At its open end the rough seas constantly eroded the newly formed rock and seawater pouring into the crater was immediately vaporized by the hot lava. Gradually the eruption subsided, leaving a land barren but waiting for something to move, to grow, just as the continents must have waited millions of years ago.

It is in Iceland itself, however, that evidence of the divergence of the earth's crustal plates can best be seen. In that northern island republic the old rocks—in both south-east and north-west—are of the same material and many millions of years old. Between them lie newer rocks, mostly less than a million years old and some of them only 10,000 years old or even less. The young rock is magma that flowed up through the cracks in the crust as the plates moved apart, and hardened. As new cracks appear more magma wells up, slowly pushing the two older parts of Iceland even farther apart.

In parts of the world where the convection currents in the mantle meet and are deflected downward, the crustal plates are *converging.* An ocean plate meeting a continental plate underrides it, and at the point of collision deep ocean trenches are formed, such as the deep Marianas Trench in the western Pacific. The underriding oceanic plate also lifts the margin of the continent upward into mountain chains. The Andes Mountains as well as the mountains of the Pacific coast of North America—particularly the Cascades Range and the Olympic Mountains—are still being built up by such plate movements.

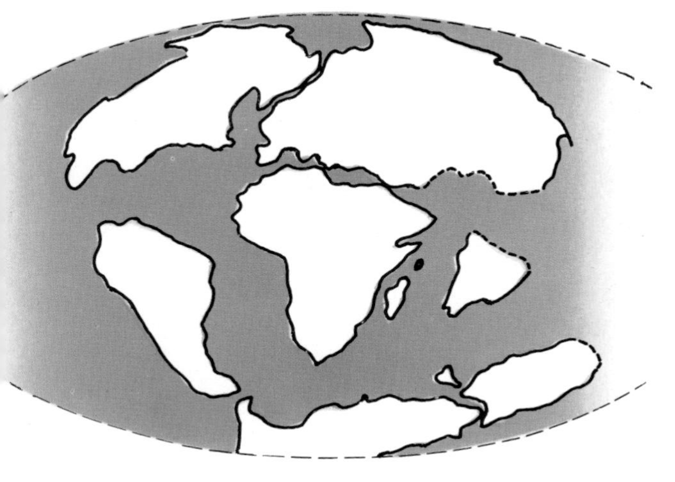

65 million years ago

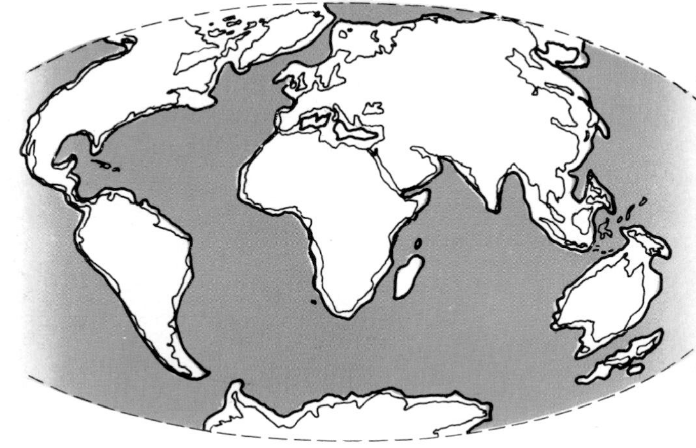

Present day

When continental plates converge, as has been happening for some millions of years as a result of India's union with Asia, the plates are pulled together with such force that the lighter upper crustal materials crumple and fold. Accumulations of sedimentary rocks in these areas, as in northern India, are squeezed outward and upward into new mountains. The great mountain ranges lying between India and the regions to the north—the Hindu Kush, the Pamirs, the Karakoram, the Himalaya, and the Kunlun Shan—are built of sedimentary rocks squeezed together in this convergence.

There are, by most reckonings, 19 crustal plates and it seems probable that their movements began as soon as the crustal materials separated out. It is also probable that as soon as discrete continental masses appeared they, too, began to drift. About 200 million years ago the plate movements had brought the continents together to form a single massive continent, Pangaea, surrounded by ocean. By fitting together all the land masses that make up our modern world, making allowance for the submerged portions and those that have been eroded and folded, it is possible to visualize Pangaea's shape. About 20 million years later Pangaea began once again to break up. First it split into two continents, Laurasia in the north and Gondwanaland in the south. Subsequently North America moved away from Laurasia, and India and Antarctica separately broke away from Gondwanaland. After another 40 million years the North Atlantic and Indian Oceans opened up. Greenland was then beginning to separate from North America, and India was on a collision course with Asia, although it had another 2000 miles to travel before they touched. The continents began to assume a more familiar shape and location about 65 million years ago. South America had by then separated from Africa, and India was nearing Asia. Australia and Antarctica were still one, as were Madagascar and Africa, although the latter were moving apart.

It is anybody's guess what the shape of the landmasses will be in another 65 million years from now. We suspect there is an upwelling convection current under East Africa and the Middle East that is causing a divergence that accounts for the great Rift Valley in Africa. The Arabian peninsula may be swinging northeastward and one day may close the Persian Gulf and push the large accumulation of sedimentary rocks in the area up into high mountain ranges.

Most of us are oblivious to these colossal movements. If, however, we lived near an active region—on an island over a divergence or in lands astride or near convergences—we could examine at first hand the volcanoes, rift valleys, and growing mountain chains that are the direct evidence of crustal activity. The majority of active volcano and earthquake centers are found along the edges of the crustal plates. But in zones no longer active, for example in Western Europe and most of the eastern and central areas of North America, the remnants of ancient volcanoes mark the sites of early plate conflicts.

Volcanoes are the most spectacular of all geological events. Although they have been responsible for the deaths of hundreds of thousands of human beings, and probably millions of other organisms, and the destruction of their

Left: molten rock jetting high in the sky below a dark cloud of ash—both emitted by an Icelandic volcano that erupted in 1973 for the first time in thousands of years. The Helgafell volcano burst open a fissure more than one mile long, almost splitting in two the tiny island of Heimaey. (Significantly, Heimaey lies only 12 miles northeast of the new volcanic island of Surtsey.)

Below: beautiful but lethal streams of lava light up the night as they cascade down a slope. Such a river of lava containing many thousands of tons of molten rock oozed two miles across the island of Heimaey and into the sea.

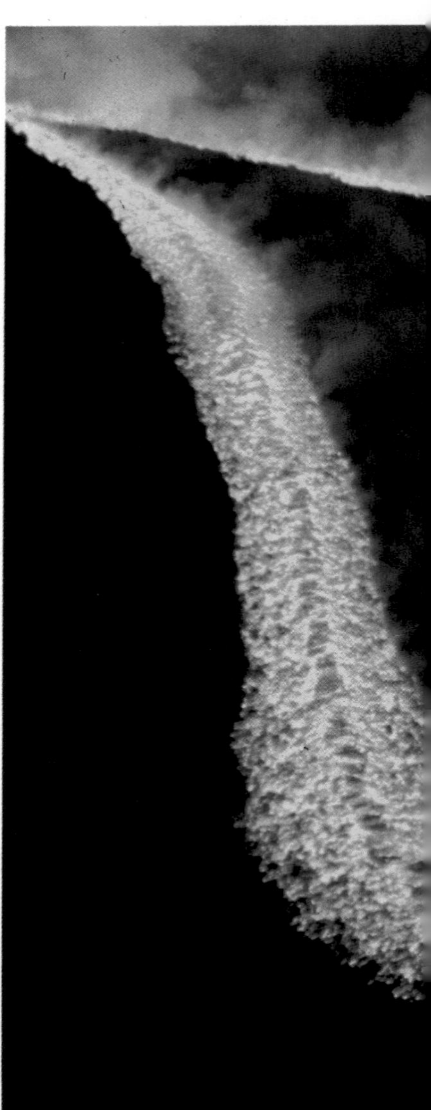

habitats, they replenish the minerals and *igneous* (heat-formed) rocks from which new sediments are formed by erosion. There are many kinds of volcano, but beneath each of them there is a chamber of magma under immense pressure. When the pressure is so great that the magma is forced to the surface, it emerges as incandescent *lava*. The distinctive cone of a classical volcano is usually built up of layers of lava and ash. A volcanic eruption produces in addition to lava and ash large volumes of water vapor, oxides of sulfur and nitrogen, and other gases.

Volcanoes may build shallow cones containing lakes of molten lava and release their gases quietly or in explosive eruptions. In some Icelandic eruptions, lava issues from long fissures, producing lava sheets that may cover hundreds of square miles. For example, the 1783 Laki eruption was a fissure eruption that produced a lava flow that spread over the island until it covered an area of some 173 square miles. Its total volume could have covered most of the New England states with a four-inch layer.

The formation of a new volcano is a fearful but wondrous sight. Early in the spring of 1943 the villagers of Paricutin, 200 miles from Mexico City, were disturbed by a series of slight earth tremors. For a few weeks the rumbles gradually increased in intensity and then, one day, gas-belching cracks appeared in a nearby cornfield. The rumbles soon increased in intensity and frequency, and explosive blasts of hot gas enlarged the cracks into craterlike holes. Lava overflowed the cornfields, and hot cinders were catapulted into the air.

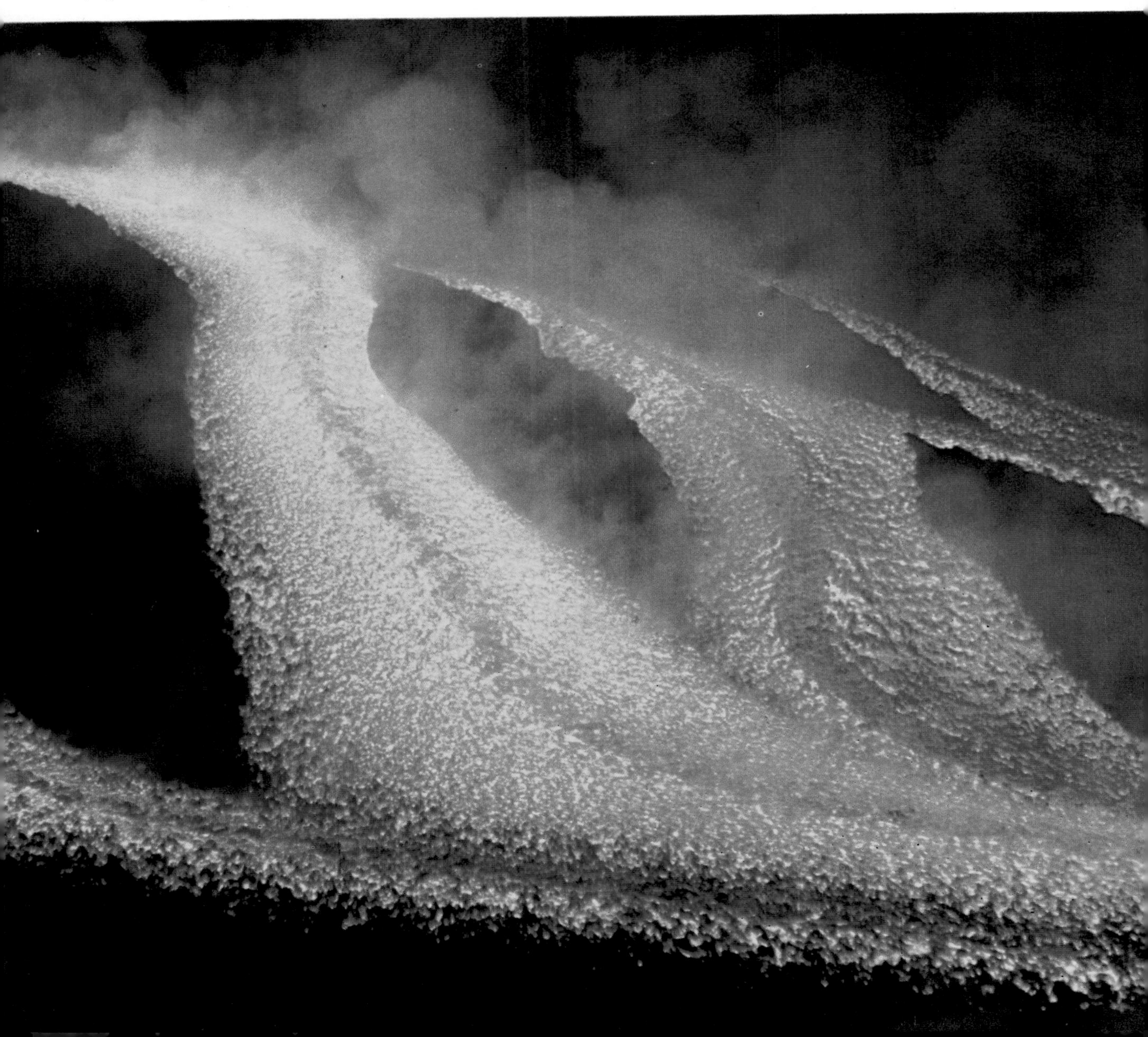

With each passing week the earth tremors and the lava and ash flow became more intense. By midsummer the villagers of Paricutin and other settlements nearby had to be evacuated. Eventually, the slow-moving but fiercely hot lava destroyed the little village. By the end of September the volcano had formed a perfect cone, 1500 feet high and nearly a mile in diameter. The volcano of Paricutin is now "dead" and the erosive forces are beginning the long task of leveling it.

Between eruptions volcanoes may lie dormant for so long that people living near them believe they are dead, and in a sense of false security build cities nearby only to see them subsequently destroyed.

The most famous of all volcanic catastrophes was the destruction in A.D. 79 of Pompeii and Herculaneum by the "dead" Vesuvius. Vesuvius remained active from time to time for another 1000 years. From about 1139 it seemed to lie dormant until 1631 when it again became active and has erupted irregularly ever since. The last major eruption was in 1944, while the Allies and Germans were struggling for possession of Italy. Other active volcanoes in the Mediterranean area are Stromboli and Vulcano in the Lipari Islands. Vulcano is the site of the forge of Vulcan in classical mythology, from which volcanoes get their name.

Europe's largest volcano is the mighty Etna, which is more than 35 miles in circumference and stands nearly 10,000 feet above the island of Sicily. Etna erupts about 15 times every 100 years but its eruptions are irregular and this makes them difficult to forecast. The most recent occurred in 1971.

In 1902, on the other side of the Atlantic, a small volcanic mountain called Pelée on the beautiful island of Martinique in the Caribbean exploded, destroying the city of Saint-Pierre. The 30,000 inhabitants of the beautiful, bustling city, known as the Paris of the West Indies, were killed within five minutes, not by lava, but by the cloud of superheated gases and glowing ash particles thrown out by the eruption.

In 1883 the greatest recorded volcanic eruption disintegrated the island of Krakatoa in the Sunda Strait between Java and Sumatra. All that remained was a crater half a mile in diameter lying under 1000 feet of water. Since the island originally rose 1000 feet above sea level, the explosion must have pulverized several cubic miles of rock. The explosive eruption of Krakatoa may have been caused by seawater that had poured into an earlier crater and cooled the lava to produce a solid cap over the magma chamber beneath. This strong cap held until the pressure below blasted it apart in the catastrophic explosion that was to result in the deaths of over 100,000 people.

For some people a volcanic eruption means more than a lively lesson in geophysics; it means the destruction of home and livelihood. Heimaey inhabitants proved particularly vulnerable. Volcanic ash crushed houses (right) and citizens abandoned the island's only town to ash and lava flow (above). Most buildings suffered damage in the months while the eruption lasted.

Old volcano cones above the permanent snow line may fill with ice. Should they become active again, the ice will suddenly melt and severe flooding may occur. The 1918 eruption of Katla under the Myrdalsjokull ice cap in Iceland poured out floodwaters estimated at three times

Three scenes on Mount Etna, Europe's largest active volcano, which rises almost 11,000 feet and dominates northeast Sicily. Left: smoke plume rising from Mount Etna in 1974. Below left: sampling gas from the foot of a lava flow during a major eruption in 1971. About 140 eruptions have been recorded for Mount Etna, the worst in 1169, 1669, and 1852. Below: here, an eruption has covered large tracts of mountainside with a thick crust of smoking cinders resembling a gigantic rural slag heap. Trees are still standing in the distance, but where cinders lie nothing survives.

the discharge of the mighty Amazon. The activity could, however, be so slight as to result only in the formation of a series of ice caves such as those in the ice-filled crater of Mount Rainier in the Cascade Mountains of Washington, where snow and ice accumulation keep pace with melting and ice caves have become a more or less permanent feature.

When a volcanic mountain collapses, either explosively or when the magma chamber retracts beneath it, the solidified lava at the top falls into the hollow left behind. Often the remnants of a once fiery mountain are marked by a rim of hardened lava enclosing a flat plain rather like a lunar crater. The most spectacular of these plains is the game-rich plain of Serengeti in Kenya. When an island volcano collapses, the sea floods the central plain to make a horseshoe-shaped island. These features of collapsed volcanoes are called *calderas*. The Aegean island of Thera, north of Crete, is a caldera that blew up several thousand years ago during the time of the Minoan civilization and is believed to be one of the possible sites of the legendary city of Atlantis.

If the pressure of upwelling magma is not quite high enough to break through the resistant rocks above, but is sufficiently high to distort them, the overlying layers may be forced upward into a great dome. When the magma cools, it will harden into a lens-shaped lump of rock called a *laccolith*. Beneath the Black Hills of Dakota lie great mushroom-shaped laccoliths some of which are a mile in thickness and several miles in diameter.

Vesuvius in southern Italy is Europe's most notoriously dangerous volcano. Above: cauliflower-shaped dust cloud produced by a major eruption in 1944. Below: casts of three bodies from Pompeii, a Roman city overwhelmed by the eruption of A.D. 79.

In some parts of the world extensive fields of solidified magma extend deep into the earth's crust. These formations, which grow thicker with depth, may cover hundreds or thousands of square miles, and are often associated with mountain regions. *Batholiths*, as they are called, underlie the Rockies and Sierra Nevadas in the United States.

If magma penetrates into vertical cracks in the surface layers and solidifies, it forms structures of hard rock known as *dikes*. They vary in width from a few inches to hundreds of feet and may be many miles long. Dikes very often can be seen as features of coastal landscapes where they have been exposed after the sea has eroded away the softer rocks around them. Sometimes, however, the dikes are formed of soft rock and tend to erode faster than the rock around them, forming *chasms*. Magma occasionally forces its way between horizontal layers of sedimentary rocks to form *sills*, some of which may be hundreds of feet thick, several miles wide, and tens of miles long.

Although volcanic activity is a frequent cause of earthquakes or tremors, these can also result from heat-caused strains in the crustal rocks. The mantle and crust are poor conductors of heat and when the large quantities of heat generated in the planet's interior reach them tremendous strains are caused by the temperature differences set up between the hot lower rocks and the cooler one nearer the surface. These strains increase to the point at which something has to give, and the rock cracks almost in the same way as a cold glass tumbler does when very hot water is poured into it. A slight adjustment, even a few inches, will involve millions of tons of rock and release an enormous amount of energy. In a fraction of a second the shock wave travels outward from the area much as ripples in a pond do when a stone is dropped into it.

Although earthquake activity may occur almost anywhere on the earth's surface, it most frequently takes place along two main fracture belts. By relieving the strains in these areas, earthquakes probably protect the rest of the world from more serious shocks. The belt of greatest activity lies along a narrow band circling the Pacific Ocean. The regions of next greatest activity occur along a line extending from Morocco north through the Mediterranean, then eastward through the Middle East to the Himalaya, through China and southwest to Indonesia where it joins the Pacific belt. Japan, in the middle of the Pacific-rim earthquake area, suffers an average of six major shocks every year, and a day rarely goes by without two or three minor shocks.

Earthquakes occurring in deep oceans also follow well-defined lines. One such line follows the mid-Atlantic ridge from Iceland down to Bouvet Island in the South Atlantic, then swings between South Africa and Antarctica, up into the Indian Ocean and the Gulf of Aden. The movement of the American continental plate westward to override the Pacific Ocean plate has produced an active earthquake belt running up the whole west coast of South America, through Central America and California to Alaska. It is

This old photograph shows the stricken city of Saint Pierre on Martinique, after the city's destruction in 1902 by volcanic Mount Pelée in the background. Gas pressure building up inside the volcano blew off its solid lava plug, and gas clouds charged with glowing lava particles stormed downhill, annihilating Saint Pierre and its 30,000 inhabitants. Such outbursts reduced the gas pressure inside the volcano, which eventually sealed itself with a new lava plug, the spine seen rising from the distant mountain.

joined at the tip of South America by an arm of the Atlantic belt.

When the crustal rocks are under great strain, they crack along *fault lines*—that is, lines of crustal weakness—and produce vertical or lateral displacements, sometimes both, in the rock layers. Cliffs, or *fault scarps*, may form along the fault lines. Once such an area has developed it may remain a source of weakness for millions of years. Vertical faulting, if it continues over a long period, may raise fault blocks to form mountains or plateaus, such as the Sierra Nevada in California. These tall mountains are backed by an eastward-ranging series of tilted blocks that underlie more than 125,000 square miles of the Great Basin that occupies the western third of Utah and all of Nevada. One of the most famous cracks in the crust, the San Andreas fault, runs through California. The city of San Francisco is built on top of it.

Most earthquakes shake the ground, and in severe shocks may topple buildings, rupture river banks, or cause crustal subsidence. Less frequent are earthquakes that actually split the ground open but they are usually more severe. One such shock that occurred in Quetta in Pakistan in 1935 completely destroyed the town. People, animals, and buildings fell into the chasms that opened up in the streets, and to add to the horror, some of the victims were trapped when the cracks partially closed again. The Chilean earthquake of 1960 caused one of the largest tears in the earth's surface. It shook an area of over 1000 square miles, causing both vertical and horizontal displacements that set the whole earth vibrating like a bell struck by a hammer. Most of the initial damage occurred in the sparsely populated mountainous parts of Chile and, considering the severity of shock, the casualties were relatively low. Even so, 500 or more people died as a direct result. The casualty list was to grow, however, because the shock traveled under the seabed and generated a shock wave in the sea that drowned over 1000 people.

The gigantic ocean waves of the kind generated by the Chilean shock are erroneously called "tidal waves." They have, however, nothing to do

The Ngorongoro Crater in northern Tanzania is a caldera—a huge depression left behind by a stupendous, explosive eruption that blasted off the cone of a volcano once sited there. This photograph shows the big lake and game-rich plains occupying the crater floor. Part of the crater rim is visible beyond.

with tides, but are massive water waves spreading outward through the ocean from the center of the shock. Scientists use the Japanese name for them, *tsunami*. Japan in particular has a bad tsunami record—it is no wonder they have a special name for them. The famous Japanese artist Hokusai immortalized one of these waves, the great wave of Kanagawa, in a print. When Krakatoa blew up, it produced a 135-foot tsunami that swamped the village of Merak over 50 miles away, and this, not the explosion itself, was the direct cause of the many deaths there. Perhaps one of the most famous, or infamous, tsunamis in history occurred during the great Lisbon earthquake of 1755 during which 40,000 Lisbonese lost their lives, half of them drowned by the surging waters.

These mountains of water, weighing millions of tons, travel through the oceans at speeds of several hundred miles an hour, and yet a tsunami running through deep water may pass unnoticed by the crew of a ship sailing over it because the surface may rise and fall only a few feet. When the tsunami reaches shallow waters it shows its size and power. Here the resistance from the shelving bottom slows the foot of the wave, and the rest of the water behind it piles up to form a huge, advancing wall that breaks over the shore destroying everything in its path. Ships lying several miles off the coast may be lifted onto the shore, even carried miles inland.

Paradoxically, despite their record of havoc, we owe to earthquakes much of what we know about the structure of the earth. The shock of breaking rock sends out a series of different types of waves. Their movements can be recorded by *seismographs*, instruments so sensitive that anyone walking nearby can set the recording pen swinging to and fro and they must therefore be housed in special buildings. The first wave to reach the seismographs is a very fast compressional wave that travels right around the earth through both solids and liquids and can be detected anywhere on its surface. These waves, called *P-waves*, have provided most of the information we have about the earth's crust, mantle, and core. The P-wave is followed by *S-waves*, which travel through solids only. S-waves are not transmitted through the core, and the inference

Oregon's Crater Lake, deepest lake in North America, occupies the caldera left 6000 years ago when Mount Mazama erupted and collapsed into a great underlying chamber. The lake's Wizard Island is a cinder cone, raised by subsequent volcanic action.

we draw is that the core must be liquid.

The study of both natural shock waves and those caused by underground nuclear explosions has shown that severe earthquakes seem to be caused by rock adjustments that may take place at any depth from near the surface down to 400 miles below it. Approximately half of them occur within 30 miles of the surface. The remaining 50 per cent originate below the 30-mile level, but none occurs between 50 and 150 miles below the surface. Seismologists observing that both P- and S-waves are slowed down in this 100-mile zone have deduced that it must be nearly molten, for if it were completely molten the S-waves would not pass through it at all. This newly discovered layer has been named the *low velocity layer*.

There is also a region below the surface where the behavior of earthquake waves has shown another discontinuity in the rocks. It was discovered by the Yugoslav geophysicist Professor Mohorovicic. While studying records of a Balkan earthquake he noted that shock waves suddenly speeded up when they reached a certain depth—about 20 miles under the continents and three miles under the oceans—and concluded that the point at which wave velocity increases marks the boundary between crust and mantle. This region, now known as the *Mohorovicic Discontinuity* or "Moho" for short, may indicate an abrupt change in chemical composition or physical characteristics between rocks of the crust and those of the mantle.

Possibly these discontinuities mark regions where the physical and chemical reactions that form the compounds of the crustal rocks are taking place. Certainly they are regions of very high temperatures and pressures. We can obtain some estimate of the temperatures from the fact that the deeper we probe into the crust the hotter it becomes, at an estimated rate of 48°c per mile. At only 60 miles down, the mantle—at a calculated temperature of more than 2700°c—must be white hot. The rate of temperature increase seems to fall off at the 60-mile depth, and we know little of the temperatures below that depth.

As might be expected, pressure also increases with depth. At about 200 miles down it reaches $1\frac{1}{2}$ million pounds per square inch—100,000 times the atmospheric pressure at sea level or 1200 times the pressure at the deepest part of the ocean.

The heat in the mantle is generated by the radioactive decay of uranium, radium, and other radioactive elements in its rocks. Even a slight rise of heat output will increase the temperature of the low-velocity layer and cause it to move slowly upward as the heat is conducted outward from the interior. When this rising layer nears the crust, it starts mountain-building activities. The pressure exerted by the overlying rocks decreases as they crack, and the nearly molten rock begins to turn liquid. Conditions are then right to start upwelling and circulation in the fluid mantle rocks. Continents will be split apart and—where there is resistance to their movement—the crust will buckle into mountains.

Two photographs show some of the consequences of earthquakes. Above: horizontal jerking movements of one part of the earth's crust in relation to another have disrupted drainage patterns along California's San Andreas Fault. Right: in the 1964 Alaskan earthquake, tremors were strong enough to heave the land into waves that tossed homes like boats on a stormy sea.

Gigantic batholiths and laccoliths will form. Volcanoes, earthquakes, and all the attendant activities of a mountain-building period will occur. When the excess internal heat has been expended, the upper mantle reverts to the near-solid phase again and the earth will experience a period of crustal calm while a new period of radioactive heat buildup starts.

The original thick sediment deposits laid down in the geosynclines are poor heat conductors and by preventing the heat in the planet's interior from escaping through to the surface may be responsible for their own uplift. For example, a buildup of heat under a sediment-filled geosyncline causes local melting and an upward movement of mantle materials, pushing the overlying sediments into large folds or *anticlines*. The sideways expansion of the moving sedimentary rocks may be restricted by the less plastic rocks around them, and as a result they fold upward into sharp corrugated folds.

The internal heat is probably not relieved at all parts of the earth's surface simultaneously. This unevenness may ensure that mountain-building activities are widely distributed, one region becoming active as another calms down. On the

This scene of earthquake-engendered disaster occurred in Shiogama city, east Japan, in 1960. A gigantic wave, triggered by a submarine earthquake near far-off Chile, plowed north through the Pacific Ocean, beaching this pleasure steamer and fishing boat and causing other havoc before it spent its force ashore.

other hand, it is possible that slow convection currents are always at work in the mantle. We do know, however, that there is sufficient energy inside the earth to keep the process going for a long time to come.

The flow of heat escaping from inside the planet is more or less the same through both the sea floor and the continents. Although there is more radioactivity in the crustal rocks than in those of the mantle and more heat is produced under the continents than under the oceans, the heat flow to the surface averages out. The reason for this may be that the low-velocity layer, where mantle rocks are at or near melting point, is closer to the surface under the oceans. In fact, this situation—the closeness of the low-velocity layer to the relatively thin crust under the oceans—may account for the upward flow of magma along the center lines of the oceans and, consequently, for the spreading of the seabed.

Whatever the mechanism may be, the materials of which the crustal rocks are made are continually being renewed. If this were to go on without a corresponding loss of old rock, the earth's crust would grow steadily thicker in proportion to the mantle. The volume of the crust and the continents nevertheless seems to remain remarkably constant, which means that the crustal and the lighter continental rocks must be disappearing at a rate similar to the rate of renewal. The reason for this seems to be that in places where downward convection currents in the mantle pull crustal plates together—that is, at convergences—some of the crustal rocks are drawn into the mantle in much the same way that clothes are pulled through an old-fashioned wringer. Once the rocks are taken into the mantle, high temperatures and pressures break up their constituent compounds and then reform them, into the same or different compounds.

The downward convection current eventually slows, turns horizontally, and continues its

A seismograph, *a sensitive device for detecting earthquakes. This instrument employs a traveling pen tracing a line across a revolving drum. Tremors produce waves in the line, thus the device can reveal strength and duration of earthquake shocks.*

Aerial view of the Simareh anticline in western Iran's Zagros Mountains. Such great upfolds in the earth's crust result when an upwelling of molten materials in the earth's mantle displaces a large mass of crustal rocks. As one mass is forced sideways against another, rock layers that had been horizontal are buckled. The consequence is a series of wavelike dips and crests.

circulation, but now in the opposite direction to its course in the upper regions. At some point, as the heat builds up again, the current begins to move upward toward the surface where it pushes against the crustal rocks before proceeding along its upper horizontal course once again. When the pressure of an upwelling current forces magma through the crust, the reformed crustal material is brought with it. At the same time it snatches a new batch of crustal rocks and carries them up too, changing their composition in the process. At points where the magma mixture actually breaks through the surface the pressure drops and many of the compounds in it solidify and line the crack, or *pipe*, with new rock. In this way the mantle refurbishes the crustal rocks as well as renewing them.

In the mantle many combinations of elements are produced, and when magma breaks through the crustal rocks, some of the compounds are deposited and concentrated in veins of minerals and metallic ores. This is the reason why many mineral lodes are found to be associated with old volcanic regions.

Water can also concentrate certain mineral compounds by dissolving them from a wide area and subsequently depositing them in one specific region. Just as water can dissolve some compounds, so, too, can molten rock. In this way

The Big Hole at Kimberley, South Africa (above), was dug for the extraction of diamonds (below). Diamonds are formed under great heat and pressure deep down in the earth and rise through volcanic funnels, in a mush of material known as kimberlite.

metal-bearing compounds are carried along with the molten rock. When it begins to cool, each compound will solidify as the temperature falls to the point at which it will crystallize. This is why metal ores and even free metals—such as copper, gold, and silver—are found concentrated in discrete bands, lodes, or seams.

Chemical action may produce one mineral from another. Diamonds, for instance, are formed when sulfur combines with the oxygen of carbon

Basalt rock in southern Iceland. Basalt is made of tiny crystals, a product of magma that cooled rapidly near the earth's surface. But the cooling also set up tensile stresses that split the rock, producing a mass of large, six-sided columns.

dioxide under the right conditions of temperature and pressure, leaving the carbon to crystallize out. These conditions are likely to be found only deep down in volcanic regions.

Rocks are constituted of differing combinations of compounds derived from about 10 of the known chemical elements. The basis of all the crust's various rocks is the *igneous rocks*, which are those formed from cooling magma. Igneous rocks formed when magma breaks out onto the crust through fissures or volcanoes are called *extrusive igneous rocks*. Pumice, obsidian, and basalt are examples. If the magma cools between layers of rocks to form laccoliths, dikes, sills, and batholiths they are known as *intrusive igneous rocks*, of which granite is the most common example.

Magmas, then, are really mixtures of melted minerals. As they cool, the minerals separate out in the form of crystals whose size and structure depend on the rate of cooling. The same magma cooling at different rates and distances from the surface will form several types of rock. Magmas that cool slowly deep within the earth's crust produce rocks with large crystals or grains such as those found in the granites. Those that cool more rapidly near the surface form rocks with tiny crystals, such as felsite and basalt. Some rocks, especially if they are formed under water, cool too rapidly for crystals to develop and are as smooth as glass—indeed, they can be described as natural glasses.

The original sedimentary rocks were made from igneous rocks by weathering and erosion. Since the original Precambrian erosion, subsequent layers of sedimentary rocks have been deposited, uplifted, and eroded many times over. Sometimes, as we have seen, they have been reformed into metamorphic rocks, which are also attacked and reduced by weathering. If they are exposed, new igneous rocks, such as those that issue from fissures and volcanoes, are immediately set upon by the forces of weathering and erosion. If, however, they are buried among the sedimentary and metamorphic rocks they remain untouched for millions of years until, by one means or another, the overlying layers are removed.

We have already seen that sedimentary rocks are formed from tiny particles in the layered sediments consolidated under the pressure exerted by the overburden. The coarser-grained sandstones are formed when the sediments are cemented together with compounds of silicon, calcium, and iron, a process speeded up by heat. Changes in environmental conditions—from wet to arid and back again—as happens in the million-year periods that are the counting units of earth history, are reflected in the nature of the sediments, which are deposited in chronological order, faithfully recording the process involved in their formation. As sediments build up on the floors of the shallow seas of the continental

shelves or in land-locked seas and lakes, the animals and plants that lived in these waters will be buried in them. The fleshy parts will decay, but the hard parts, the shells or bones, may remain as fossils incorporated in the different layers of sedimentary rock.

Metamorphic rocks are sedimentary rocks that have been altered, or metamorphosed, by the action of heat, pressure, and chemical reaction. The pressure may be brought about through the movement of the crustal rocks, when they fold during mountain building or, as we saw earlier, near volcanic regions.

Volcanoes, although they throw out mantle materials and replenish the crustal rock, unfortunately do not provide us with information about the mantle rocks, because in the process of melting and settling out as crystals the compounds formed go through several changes. Lava from a young volcano often has a different composition from that of an old volcano in the same chain.

Although volcanic regions are frequently threatening and dangerous, they are often areas of great beauty. Whether they are characterized by volcanic cones or ranges of conic mountains,

or are simply massive plateaus of lava, they are among the most fascinating and colorful landscapes on earth. Iceland, dominated by its volcanoes and ice caps, is truly a "land of ice and fire." Most of it is composed of lava sheets, now weathered into grotesque forms, covered in lichens, and the home of many flowers and birds.

Under the onslaught of weathering, volcanic rocks do in time break down to form rich soils. Iceland lies in a latitude where the soil-forming processes are slow to produce mature soils. Nevertheless, in crannies and on the open plains that have not been overgrazed, rich pasture is found. Crystal-clear lakes nestle in hollows in the lava fields, many of them having a constant temperature throughout the year because their waters come from deep within the crust, an indication of the inferno beneath. Heated rocks may, however, lie only inches under the ground surface, and in those areas mud pits grumble and bubble lethargically like hot porridge. There, too, are found clear basins of hot sulfur water rimmed with yellow crystals. Many Icelandic landscapes feature geysers blowing superheated steam out of the rocks. Geothermal energy—that is, earth heat—may be a boon. In the small town of Hveragerdi, the townsfolk grow fruit and flowers in hothouses warmed by it, and the capital, Reykjavik, is heated with steam provided by the hot rocks of that northern land.

Everything in its place, the old saying goes, and this is true of the planet earth, with its cycles within cycles, all dependent upon the tectonic movements in the earth's interior. The energy from within the earth moves the mantle rocks in a cycle that ensures a renewal of the crustal rocks. On the continents, the sun-powered forces of weathering and erosion form a second cycle of renewal intermeshed with the first. In a grand cycle that takes hundreds of millions of years to complete, the elements of which the planet is composed are separated, sorted, and recombined. The elements that make the earth's crust also compose the living material that plays its part in the crustal development. All must wait on the slow turn of the wheel of time.

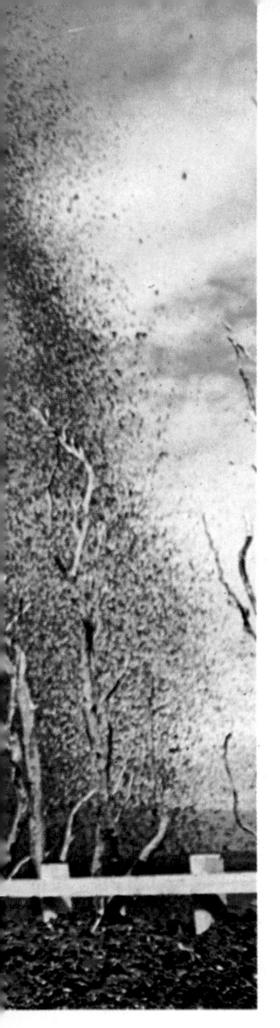

Left: in some parts of the world, volcanic eruptions frequently add rocks to the earth's surface. Here, Crater Rim Road in Hawaii Volcanoes National Park is in the course of being smothered by pumice and ash ejected from Kilauea—the world's largest active volcanic crater.

Geologically youthful Iceland largely consists of lava flows emitted no earlier than the Tertiary period. Right: such lichen-covered lava rock occupies large parts of Iceland as yet unclothed in soil. Below: steam rising near Reykjanes in southwest Iceland, one of many areas in Iceland where man taps the intense subterranean heat that often goes with volcanism. Such geothermal energy heats thousands of Icelandic homes.

The Dynamic Quartet

In the earliest of days, the earth, drenched by solar radiation and pulsating with interior heat, was the scene of a battle between three major environmental factors—the atmosphere, the hydrosphere, and the crustal rocks or *lithosphere*. It was truly a conflict among giants and was inevitably settled by a compromise of continuing balance that has lasted into our own times. In the midst of that turbulent primeval battle, a fourth environmental factor, weak and vulnerable, was stirring in the primitive oceans. As the first rocks cooled and the first oceans formed, life was passing through its most difficult stages of development, but it was evolving. The chemical forerunners of life, the self-replicating or reproducing molecules, were establishing themselves in the oceans and competing for survival.

During that long period of evolution there were probably thousands of different chemicals able to make replicas of themselves, but as conditions on the earth changed through the cooling of the crust and oceans, their numbers diminished. Those unable to continue their chemical activity became extinct, whereas those that could adapt to the changing conditions survived and their progeny carried on into the future. Eventually a time came when the survivors competed for the materials from which they could make reproductions of themselves, thus laying down an unassailable law of life still operating in all organisms from amoeba to man—the Darwinian principle of the survival of the most adaptable and the fittest.

The story of life began at the very beginning of all creation. The chemical elements of which all living organisms are made, the same elements that are the building bricks of the galaxies, stars, and planets as well as the rocks of the earth, appeared at the moment the universe was born. The conditions necessary for the creation of simple living systems were right at a very early stage in the earth's history. The chemical reactions taking place as the earth's crust formed were responsible for producing the first hydrocarbons, and in their turn these compounds led to the synthesis of other and more complicated organic materials from which life could evolve. This was the epoch of the great

Wispy cloud half hides the sun, as vapors and clouds may have blurred its image in earth's infancy. Even so, thousands of millions of years ago, solar energy helped to produce the chemical changes that gave rise to the first living cells. Later, ozone gas derived from the oxygen produced by early green cells stopped lethal rays from the sun reaching the earth. After this, advanced life could evolve.

Fossil impression of one of the earliest known creatures, a primitive organism that lived in Precambrian times, more than 600 million years ago. Found in South Australia, and grandly named Tribrachidium heraldicum, *it lacks any known relatives.*

synthesis, the time of the "primitive soup," when water in its liquid phase became plentiful and the planet's oceans were forming under a violent atmosphere of methane, carbon dioxide, water vapor, and ammonia.

The very first stage in life's long story may have been the synthesis of amino acids from this primitive atmosphere. The atmospheric gases provided the requisite elements carbon, hydrogen, nitrogen, and oxygen. The energy to rearrange them into new compounds could possibly have been supplied both by the lightning discharges that are thought to have been common and by the intense, unshielded radiation from the sun. Once amino acids had been synthesized, it was a fairly short step to the production of proteins, which are composed of linked amino acids. The next advance was the chemical evolution of nucleic acids.

Once these compounds—amino acids, proteins, and nucleic acids—appeared in the oceans' nutrient broth, life had arrived on planet earth. At some point the nucleic acids reached the stage where they could utilize the proteins to organize duplicates of themselves; thus self-replication, or reproduction, was established, and the continuation and further development of life was assured.

It is possible that mixtures of proteins and nucleic acids joined in chemically cooperative relationships were able to surround themselves with a bubble of their environment. In such tiny parcels they were protected from the outside environmental changes that were becoming more frequent. Either in this way or by some similar means something resembling primitive living cells appeared, to wash around in the seas absorbing the free amino acids and proteins.

It is likely that these first cellular organisms reproduced by dividing in two, a method found in bacteria and other simple forms of life today. In this form of reproduction cells simply divide into two replicas of the original. Each of these then divides, and so on for generations. Provided that the environment is favorable and supplied with sufficient nutrients, a single bacterial cell can produce millions of descendants in a few hours.

The chemistry of reproduction, however, while conforming to the grand pattern of organization, is sometimes inexact in following details. As generation follows generation small differences appear, some of which, if they are advantageous, are passed on to succeeding generations of cells. In time, the result of accumulated small errors in following the original reproductive pattern is a new cell—or organism—quite different from the original ancestor. The primitive oceans contained an abundance of nutrient chemicals derived from the erosion of the first continental rocks and dissolved from the atmosphere, which were sufficient to support vast populations of simple organisms. And in 1000 million years or so, there was time enough for many variants to appear. During those early times, a wide variety of primitive cells must have evolved. Many of them died out, but the most vigorous survived and adapted to the changes in the environment. The most successful of all were cells that had evolved the capability of synthesizing the green pigment *chlorophyll.*

Cells with chlorophyll can trap the energy of sunlight and use it to reorganize molecules of water and carbon dioxide into the chemicals necessary for life and to release the excess oxygen from the reaction into the environment. When

Trilobite fossil from Vermont. Some 10,000 species of these now-extinct sowbuglike creatures once foraged on the seabed. Evolving from some kind of worm, they appeared in the fossil record 600 million years ago, and survived 375 million years.

3200 million years ago: earliest primitive unicellular forms of life.

Later (from 2800 million years ago); algae (stromatolites). 680 million years ago jellyfishes, etc.

Sudden appearance of many invertebrate groups.

Dominance of trilobites in seas.

Graptolites, trilobites, and brachiopods abundant.

Primitive jawless fishes. Rich coral reef faunas. First land plants.

Bony fishes, including air-breathing forms.

Rise of amphibians and insects.

Abundant marine faunas.

Spread of reptiles and insects. Many extinctions in shelf sea fauna

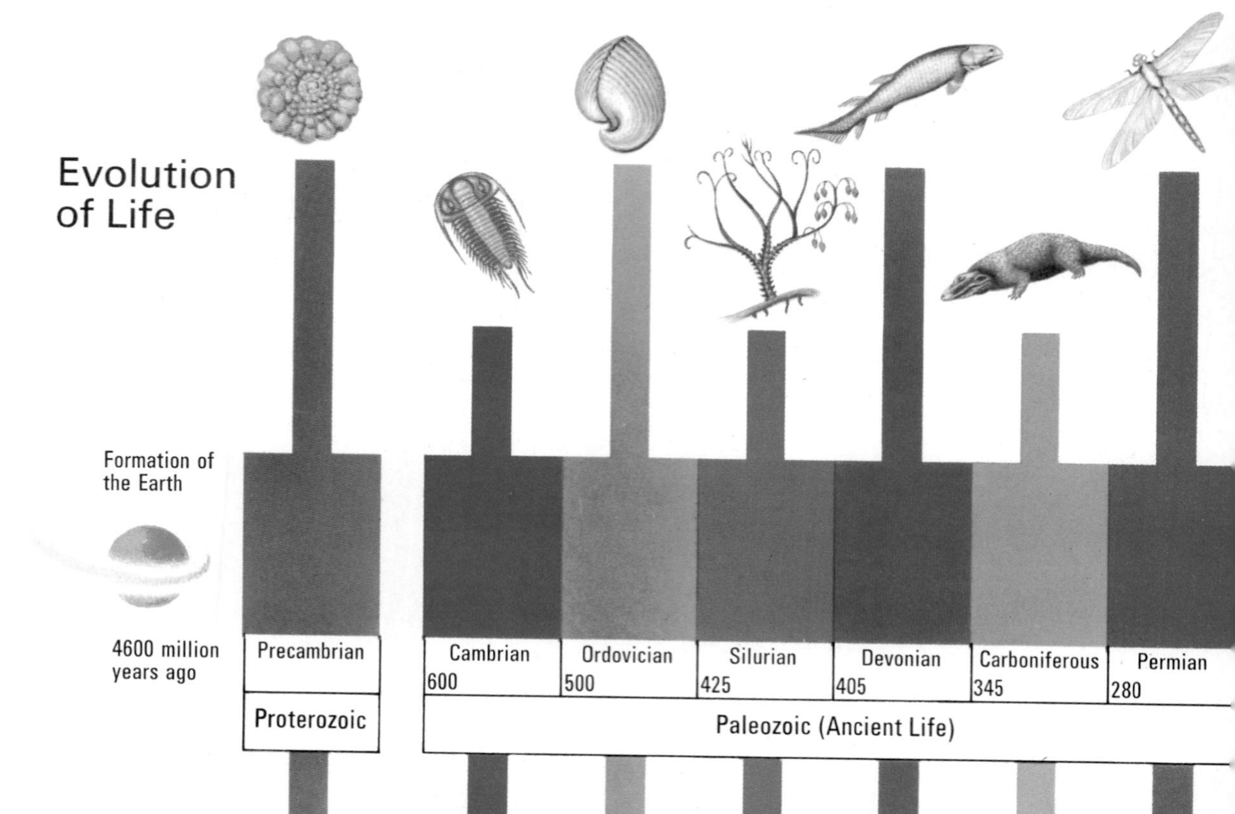

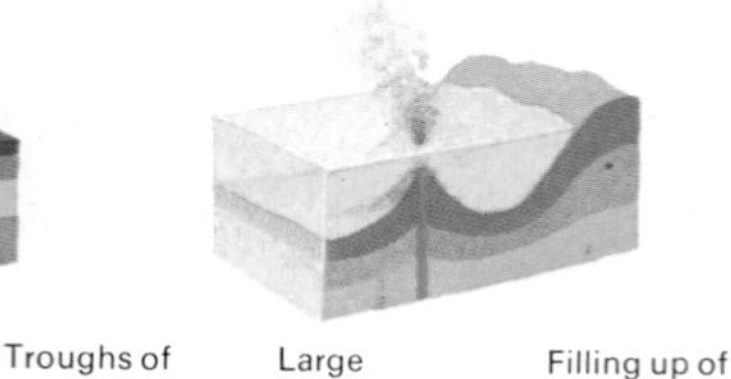

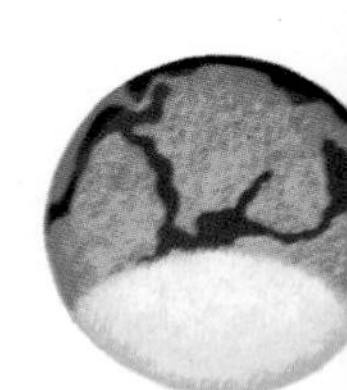

Formation of crust. Episodes of mountain-building. Igneous activity.

Troughs of sedimentation formed.

Large thicknesses of sediments deposited in troughs. Volcanoes on land and under sea.

Filling up of troughs. Growth of reefs.

Caledonian earth movements. Mountain-building and igneous activity.

Tropical coal-forming forests in coastal swamps.

Ice Age over much of Southern Hemisphere. In north, Hercynian earth movements form mountain chains.

Landmarks in the evolution of life (above) and of the earth's crust (below) related to the eras, periods, and epochs of earth history. Precambrian time (the first phase) is sometimes divided into three successive eras ending with the Proterozoic (earlier life). The next two eras (Paleozoic and Mesozoic) contain the periods Cambrian through Cretaceous. The Cenozoic era comprises the Tertiary

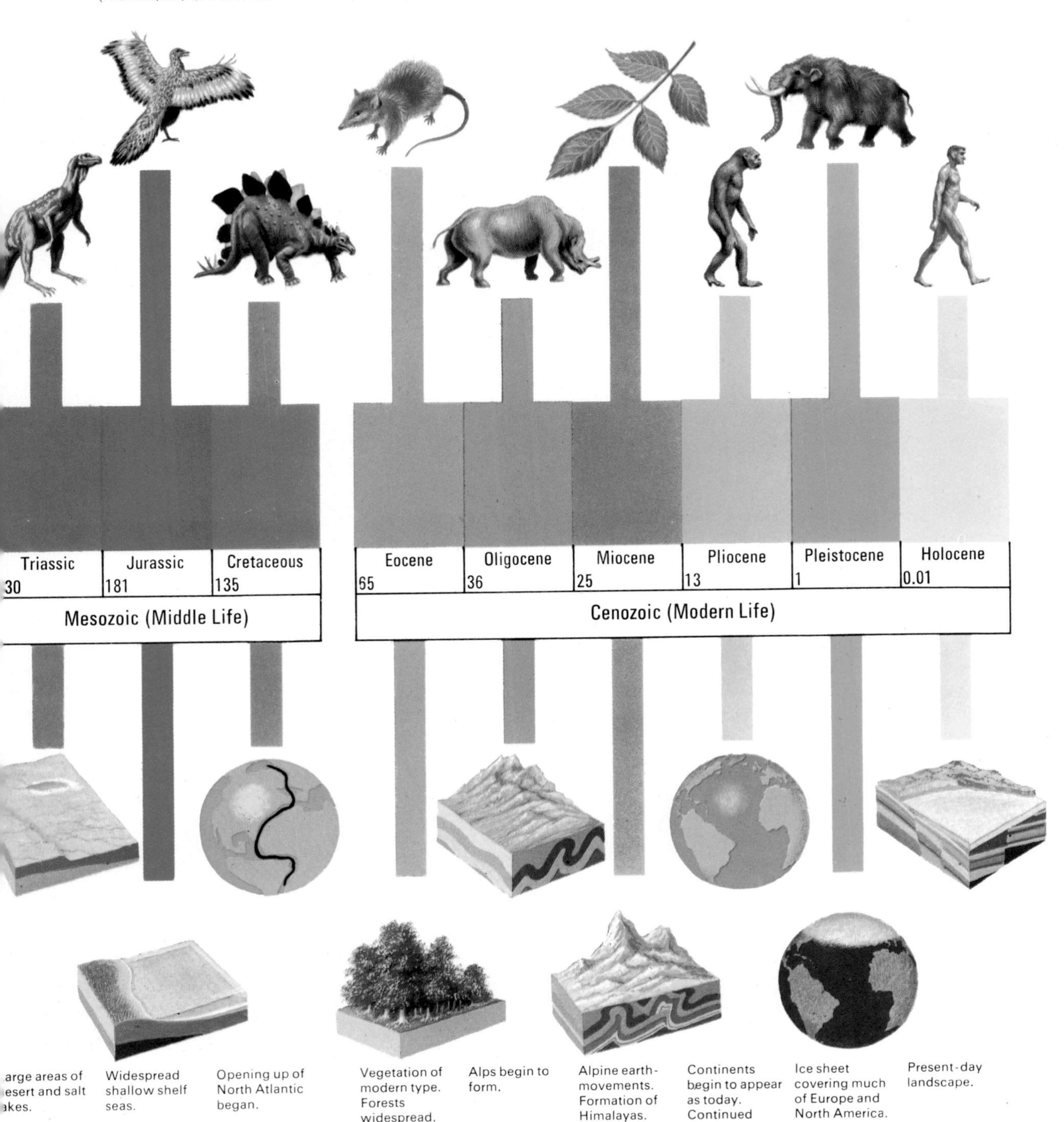

period (about 65–1 million years ago) and the Quaternary period (the last million years), subdivided into epochs as shown. Many authorities divide the time span 65–35 million years ago into two epochs: Paleocene (65–55 million years ago) and Eocene (55–35 million years ago). Some experts put the beginning of the Pleistocene epoch more than a million years ago.

Aerial view of part of the coral reef fringing volcanic Bora-Bora Island in the Pacific Ocean. Because the tiny coral-reef-building organisms thrive in shallow water, they tend to build the coral upward to keep pace with the island's slow submergence. In such ways living things have added substantial thicknesses of coral limestone to the sedimentary rocks of the the earth's crust.

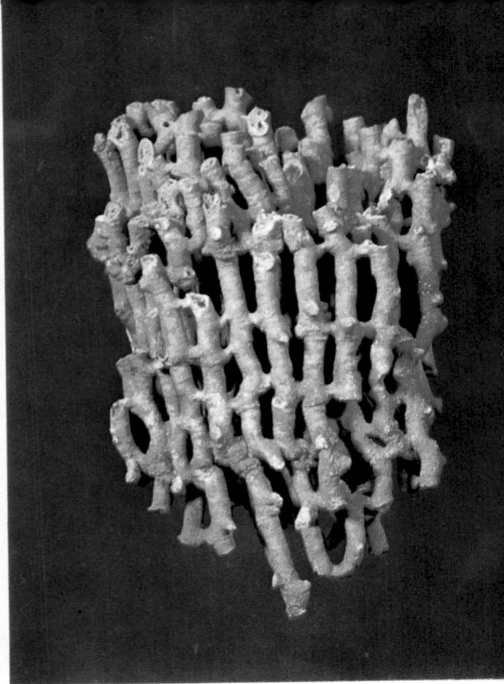

Corals have remained basically the same since Paleozoic times. These living corals (left) and this fossil coral (right) were both built from limestone skeletons secreted by tiny marine organisms. Coral polyps encase their vulnerable bodies in tube-shaped limestone shells, which they extend upward as they grow. Different growth patterns help to produce different coral shapes.

the first primitive green cells came into being between 1000 and 2000 million years ago, they began the final change necessary for the production of free oxygen in the planet's atmosphere and the establishment of a complex cycling of carbon from the nonliving environment into living organisms and back again.

The first single-celled green organisms, having evolved in a world of intense sunlight, were adapted to utilize abundant energy and were therefore vulnerable to reductions in light intensity. During the frequent periods of volcanic activity in this early stage of the earth's crust clouds of volcanic ash and gases cut the incoming solar radiation and cells dependent on intense sunlight died off, leaving those that could survive on a smaller radiation input to evolve.

The free oxygen released into the atmosphere first oxidized the minerals in the continental rocks and then began to replace the gases of the primeval atmosphere until, about 1000 million years ago, the proportion of free oxygen in the atmosphere reached one per cent. At this critical point in the earth's history there was sufficient free oxygen to reach the upper layers of the atmospheric envelope where radiation from the sun and deep space changed it into another form of gaseous oxygen called *ozone*. This gas forms a protective shield against most of the lethal radiations. Once the protective ozone layer was established the photosynthesizing cells were able to live nearer the surface of the oceans and lakes and shallow waters, where they competed with the remaining prebiotic molecules and were probably responsible for their extinction by depriving them of the chemical nutrients necessary for their "living" functions.

The evolution of single cells into multicelled organisms took a long time to accomplish. It probably happened by accident, an occurrence that in other circumstances might have spelled doom to a developing organism. Possibly a cell in the process of dividing could not complete the division and the two stayed joined together in the form of a two-celled, dumbbell-shaped organism that may have been the ancestor of all multicelled organisms. The separation mechanism failed again and again as the cells produced instead of daughter cells a long chain of connected cells. Perhaps the separation mechanism then came into play again, allowing the production of daughter cells at either end of the chain. In time there would be a whole variety of these long thin strands of life. It seems certain that some came into contact with the bottom sediments of the shallow seas and stuck there, held by the cell wall. Alternatively, a single cell may have stuck fast and then gave rise to long thin tendril-like plants that reproduced by budding. Many computations of how this may have happened can be made and all are possible.

These strands of cells slowly evolved into the first algae-like plants, a group of plants that have successfully come down to our own era. In time, the rocks and sediments were covered by a green slimy carpet of single and multicelled plants, and eventually the first seaweeds appeared. The mechanical effects of these early plantlike growths were considerable, and foreshadowed the effects of their descendants; their fronds tended to slow down the water movement, thus persuading it to drop its suspended sediments. The obstruction formed by the plants in any case tended to collect the sediments around them.

Although photosynthesizing organisms were the most numerous of living things, they were not the only ones. All cells did not, and still do not, obtain energy through photosynthesis. Some had evolved methods of obtaining energy by means of chemical processes that produced iron and sulfur instead of oxygen. Similar life processes are still found among bacteria that obtain energy by breaking down iron and sulfur compounds.

The first simple organisms were the dominant forms of life for tens, possibly hundreds, of millions of years. All through this settling-in period, life was limited to the oceans and seas of the continental shelves, and for most of that long time span consisted mostly of the algae-like plants, into which the green cells had evolved. But very slowly, new developments were taking place. A group of cells unable to provide their own energy attached themselves to the photosynthesizing plants and extracted ready-made amino acids, proteins, carbohydrates, and fats. These were the first herbivorous cells, the ancestors of the evolutionary line that developed into animals. Some were probably able to move around, swimming from food plant to food plant. Once established, mobility opened up all sorts of evolutionary possibilities. The first requirement for this new line of evolution was abundant free energy and this the free oxygen provided.

In the seas, the animal-like cells began to resemble the familiar life forms of today. All the major groups of animals were to descend from these early proto-animals and many of the earliest forms are still represented in the oceans, an indication of the stability of marine environments. Quite early in their evolution, some of the animals developed the ability of storing calcium

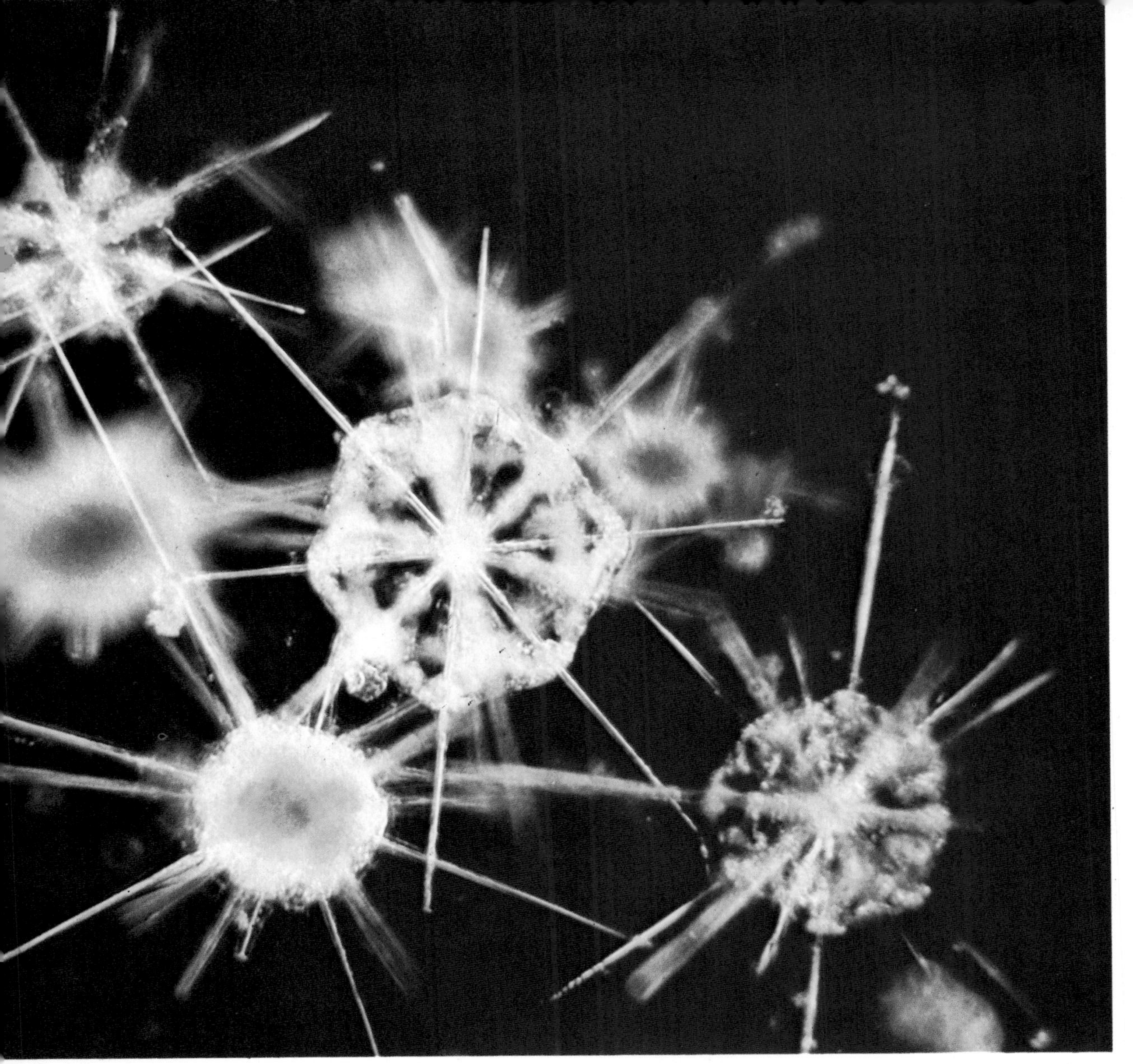

Above: an enlarged photograph of living radiolarians—tiny, colorful, delicate, spiky sea creatures, many with skeletons rich in silica. Right: shells of dead radiolarians found upon the ocean floor. Since Paleozoic times, countless millions of dead radiolarians have drifted down to the seabed, creating oozes that have been compacted and now form sedimentary rock.

Left: magnified foraminiferan microfossils. Vast quantities of the calcium carbonate shells of these minute sea organisms, known from early Paleozoic times, form marine oozes. They also occur in rocks under more than two thirds of all present land.

Above: fossil graptolites on shale. These tiny sea creatures with chitinous skeletons evolved during the Ordovician period. Below: fossil crinoid or "sea lily," member of another group of simple animals that became common in Ordovician seas. Their hard parts helped to form the so-called crinoidal limestones.

and silicon, which gave the cells a certain rigidity. In time, many storage systems became structural supports or protective armor. Thus the beginnings of skeletal structures appeared.

The selection and concentration of several elements had by now become an important part of life. When these organisms died their shells and skeletons formed vast sediment deposits of highly concentrated compounds of calcium and silicon in combination with carbon and oxygen. In time these deposits were converted into sedimentary rocks, locking up the elements they contained for millions of years until tectonic forces pushed them up above the oceans, to be released for recycling by weathering and erosion.

Life had now joined the hydrosphere, the atmosphere, and the lithosphere to form the dynamic quartet of the earth's crustal story. First, the released oxygen produced oxides of the elements in the rocks and altered their rate of weathering and erosion. Then there was the selection, storage, and concentration of elements so that the old materials of life would in due time appear as new rock.

At this point we have covered some 90 per cent of the earth's total history—that is, total history to this moment. We have pictured the story of life in broad strokes because, although we know generally what must have happened, the details have been so obscured by the movements of the crust and the inexorable forces of erosion that the record of geological and biological events remains imprecise. Those 4000 million or more years must be regarded as a single dynamic unit, known as Precambrian time.

Once life had developed to the point where organisms could leave a useful fossil record in the layered sediments, it became possible to divide the earth's subsequent history into geological divisions (called *eras*), each marked by quite distinct shifts in the general character of life and in the dominant species.

The eras are the *Paleozoic* (ancient life), which began 600 million years ago; the *Mesozoic* (middle life) beginning 230 million years ago; and the *Cenozoic* (recent life) beginning 63 million years ago. Each era has been subdivided into periods, again based on the dominant life forms recorded in the fossil record. The periods are named either for the area where their typical rocks were first identified or for some characteristic of the rocks.

There are six Paleozoic periods: the *Cambrian*, after the ancient Roman name for Wales; the

Ordovician and *Silurian* periods, after territories of ancient Welsh tribes; the *Devonian*, from the English county of Devonshire; the *Carboniferous*, or coal-bearing period; and the *Permian* after the ancient Russian kingdom of Perm. In the Mesozoic era, the first period, *Triassic*, tells us that its typical rocks were triply folded. The *Jurassic*, second period, rocks were studied in the Jura Mountains of France, and the third period, the *Cretaceous*, was named for its chalky rocks.

Our own era, the Cenozoic, has had two periods, the Tertiary, and the Quaternary; the first of these is further subdivided into five epochs. These epochs are named on the basis of groups of organisms surviving from the beginning to the present date. The divisions are the *Recent*, the *Pleistocene* (most recent), *Pliocene* (more recent), *Miocene* (less recent), *Oligocene* (few recent), *Eocene* (dawn recent), and *Paleocene* (ancient recent).

It was not until the end of the Precambrian and into the Cambrian that many of the groups of marine plants and animals that were to survive the succeeding 600 million years appeared in the world's oceans. During the 100 million years of the Cambrian, life advanced in complexity. By the time the period ended, half the evolutionary story had been written.

The fossil record breaks sharply between the Precambrian and the Cambrian. Fossils are rare in the rocks laid down in the former, whereas there is a profusion of fossils in the rocks of the latter. One reason for the rarity of Precambrian fossils is that most of its organisms were soft-bodied and, except for a few rare organisms, left no fossil trace. Another reason is that a long period of erosion toward the end of the Precambrian destroyed much of the fossil evidence.

By the middle of the Cambrian, the Precambrian organisms that developed shells and armor had blossomed into their prime. The evolutionary pressure to develop armor for protection may have been triggered by the increase in their population. As the population rose there would be increasing competition for the fairly stable food supply that existed in the conditions of the time, and shells and armor would have a decided competitive advantage. The protective coverings were either of chitin, the hard material that forms the outer skeletons of insects, or shells of calcium carbonate such as we find in the mollusks. Because this development made it likely that some of the armored organisms would

Above: fossil cystoid found in Ordovician rocks in southeastern Canada. This extinct type of marine creature was related to the modern starfish and sea urchin. Below: living cuttlefish, one of the cephalopod group that includes the squid and octopus. Cephalopods evolved more than 500 million years ago.

Left: fossil eurypterid from Silurian rocks in western New York State. Such now-extinct creatures were arthropods—relatives of trilobites and spiders. Their nickname "sea scorpions" derives from their underwater mode of life and their scorpionlike bodies armed with pincers. Eurypterids prowled the beds of lakes and lagoons in search of prey.

Right: Birkenia, *a small, freshwater type of fish widely distributed by late Silurian times. Like other early fishes,* Birkenia *lacked true jaws. But it probably swam more actively than those that were enveloped in bony armor. In spite of this defense, such slow-moving creatures were doubtless preyed upon, possibly wiped out, by the eurypterids.*

be preserved in the sediments, we have a fairly clear picture of the evolutionary lines taken by organisms from the beginning of the Cambrian period to a few geological moments ago. The evolutionary advantages of possessing a shell also ensured that in the course of time carbon in the form of carbonates—an essential ingredient of the limey shells of many marine animals—would be extracted from the environment and deposited in vast layers of carbonate sediments in the seas and oceans, as countless millions of organisms died over tens of millions of years.

Early in the Cambrian period the seas began to flood large areas of the earth's surface that had been worn down to sea level by Precambrian erosion. The flooding was increased by a general rise in sea level caused in part by the melting glaciers and ice caps. Although the Cambrian had started cool, the climate later moderated and an earth-wide summer began that was to last for millions of years.

Cambrian North America consisted of a flat landmass surrounded by sea, but to the west, south, and east lay three offshore mountainous island areas: Appalachia, stretching from Florida to Newfoundland; Llanoria, where Texas and New Mexico lie today; and Cascadia to the west. The sediments eroded from the rocks of these mountainous islands settled out in the geosynclines lying between them and the continental landmass. On two occasions during the Cambrian the seas advanced and retreated, but the crust was generally stable. Lands rose gently out of the sea and were slowly weathered and eroded away. As the Cambrian period drifted to a close, the seas once again retreated, leaving North America as a low-flanked continent, and the land in Europe began to rise. Although it is possible that the first intertidal plants and animals appeared at this time, the Cambrian continents were otherwise desolate. To us, the Cambrian landscape would have presented an extremely alien picture, for in our times there is hardly a place, even in the deserts, where life of some sort does not exist.

Dominance by one animal group or other has been a feature of the story of life. The first dominant groups, the *trilobites*, appeared in the Cambrian seas and survived for 370 million years. The segmented bodies of these armored creatures resembled sow bugs or wood lice and varied in length from one inch or so to an 18-inch species that weighed eight to 10 pounds. The dominance and long survival of the trilobites can be attributed to several evolutionary developments. They lived both by scavenging on the remains of other organisms—the most abundant source of food—and by active predation; they were highly organized animals with well-developed nervous systems and light sensors, and were capable of coordinated leg movements. A further advantage was that they could curl up in their armor when attacked. These traits gave trilobites an advantage in finding food and saved them from becoming food themselves. But, above all, they were immensely adaptable. To judge from their fossil remains the trilobites were certainly numerous, for they constituted about 60 per cent of animal life in the Cambrian.

Next most numerous were the rather mollusk-like *brachiopods*, of which some 200 species are

The Devonian Period

1 Duisbergia
2 Psilophyton
3 Pseudobornia
4 Horsetails
5 Nematophyton
6 Lycopods and ferns
7 Rhynia and Hornea
8 Pteraspis
9 Pseudosporochnus
10 Crinoids
11 Coral
12 Lampshells
13 Cephalaspis

Imaginary landscape of Devonian times. Primitive plants (1–4, 6, 7) now colonized dry land, as did early arthropods and vertebrates. But animal life was still concentrated in the seas and rivers. The composite underwater scene, below left, shows two primitive freshwater fishes (Pteraspis *and* Cephalaspis) *in a sea setting featuring crinoids, coral, and lamp shells.*

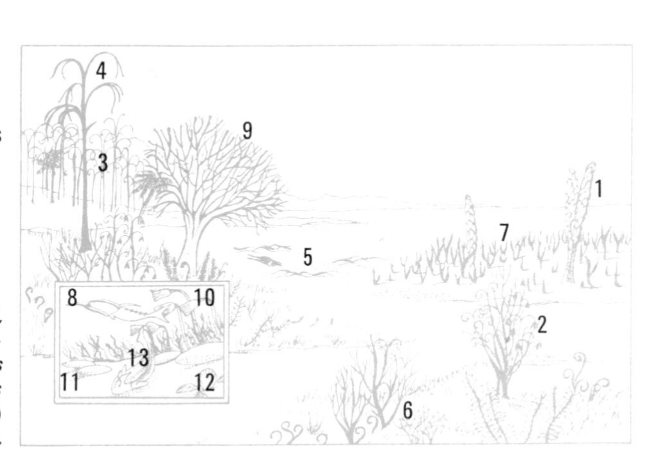

still with us today. The Cambrian seas also abounded in sponges, worms, mollusks, jellyfish, and a primitive ancestor of coral. Corals are an indicator that climates were warm, for warm water was a prime requirement for their success in the past as it is today. Corals may be solitary, in which case they form a simple cup of calcium carbonate, or colonial, when millions of them lay down communal structures. Often, the colonial corals form carbonate reefs off islands or mainland shores, such as the Great Barrier Reef off the east coast of Australia.

Reef-building corals live in shallow coastal waters because they need light as well as warmth. Around the shore of a volcanic island coral grows upward to keep at the right depth to obtain its food from the surface waters. When the island sinks, as all volcanic islands eventually do, the coral reef will be all that is left to mark its submergence. At this stage the coral reef forms an atoll, a ringlike island with a lagoon at its center where the volcanic peak once existed. Over the centuries the debris eroded from the coral will fill the lagoon until a low island is formed. Eventually the island will sink below the waves altogether, and when this happens the coral dies. All that remains is a round, flat-topped underwater mountain, called a *guyot*.

Two groups of marine organisms that appeared toward the end of the Cambrian and were to become important factors in the balance of carbon and silicon in the oceans, were the *foraminifera* and the *radiolaria*. These tiny, one-celled animals build shells of calcium carbonate and silicon dioxide respectively. When they die, their shells drop to the seabed to form, with the remains of other organisms, thick sediments or oozes. The pressure of these sediments accumulating over several million years gradually changes the lower layers into bands of sedimentary rock containing millions of tons of calcium carbonate and silicon dioxide. Thus the foraminifera and radiolaria became a vital factor in the cycling and recycling of calcium, silicon, carbon, and oxygen between the lithosphere and biosphere.

Among the Cambrian invertebrates were a few predatory animals that had developed a system for storing calcium phosphate to provide the phosphorus needed to drive their muscles. The phosphates were stored in platelike structures in the animals' skins, which not only provided a supply of phosphorus for muscle activity during times of phosphorus shortage,

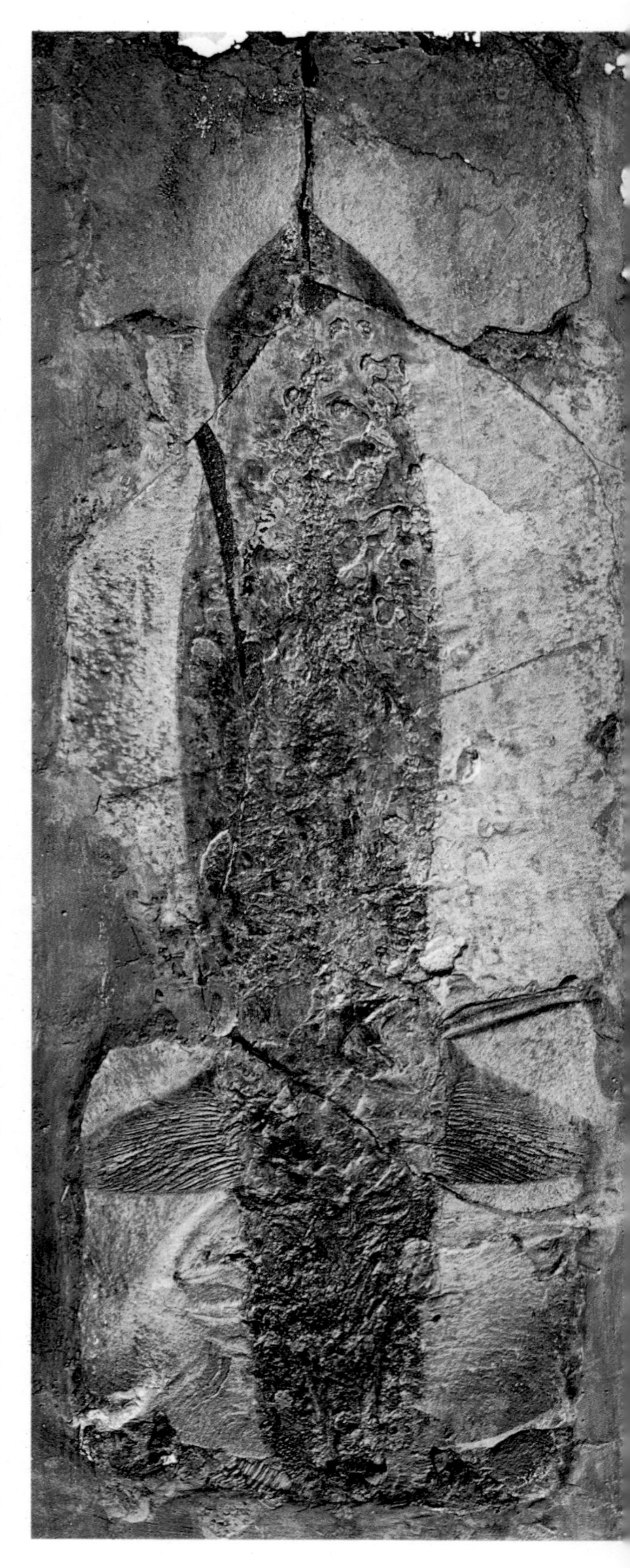

Two fishes that flourished in Devonian times, often called the "Age of Fishes." Left: Cladoselache, *a primitive shark about three feet long. This fossil specimen was embedded in shale in Ohio. Like modern sharks,* Cladoselache *possessed a skeleton of cartilage, not bone. But like most ancient fishes, the earliest sharks lived in fresh water. Right: artist's reconstruction of* Climatius, *a three-inch-long member of the now-extinct acanthodians (perhaps the first jawed fishes), misnamed "spiny sharks." Unlike a shark,* Climatius *had a covering of true scales and some developed bony structures, features found in the higher fishes of today.*

but also acted as structural supports for muscles. The plates also gave a certain amount of protection from enemies. These animals were the early ancestors of the vertebrates.

By the beginning of the Ordovician period, the second of the Paleozoic periods, several animal groups had evolved armor, secreting a hard cover either as a direct by-product of their metabolism or, as in the case of the ancestral vertebrates, as the side result of a useful store of muscle fuel. By the Ordovician, life was involved in the fate of several of the crustal elements; to hydrogen, oxygen, carbon, and nitrogen had been added iron, calcium, silicon, phosphorus, and sulfur.

The Ordovician, too, was a period of worldwide summer and it seems unlikely that ice existed anywhere on the planet during the whole of its 75-million-year span. The atmosphere and the hydrosphere were quiet and the earth was bathed in sunshine. Throughout the first 35 million years or so the land gently lifted and subsided and the seas invaded and retreated as they had done during the previous period. There was little erosion and few land sediments reached the oceans to be processed into rock. Instead, most of the rocks formed in the Ordovician were of limestone—that is, calcium carbonate—deposited by marine organisms. Halfway through the Ordovician, rising convection currents in the mantle sent rumbles and shudders rippling through the rocks, breaking the summer quiet as the continents began to move. Land was rising from beneath the shallow seas, and in North America Appalachia began to rise again after being eroded away in the Cambrian. Volcanic activity broke out in the Alaskan area of Cascadia, and on the European continental plate, mountains began to fold and buckle in Wales, England, and Scotland.

Many new organisms appeared during the Ordovician, showing life's increasing tendency to diversify. Among them were the *graptolites*, which were colonies of cooperating animals in which each had its own compartment in the chitinous body of the colony. Another group of animals that made a first appearance in the Ordovician and diversified rapidly were the *crinoids*, or sea lilies. They look like long-stalked underwater flowers, but are related to the sea urchins and starfish. The most successful of the animals to come on the scene at this time were the *cephalopods*, ancestors of squid and octopus. The cephalopods, whose name means head-footed animals because their tentacles surround their mouths, are free-swimming mollusks and, for invertebrates, have highly developed nervous systems and correspondingly elaborate behavior patterns. Descendants of the Ordovician cephalopods were to rise to a position of dominance in the Permian seas and their continuing abundance makes them an important link in the marine food chains of today.

Meanwhile, the first backboned animals were rapidly evolving into more complex forms and beginning to show signs of their successes to come. These Ordovician vertebrates were fishlike in appearance. With their armored heads and bodies, naked rear parts and tails, they looked primitive. Nevertheless, they had organs of balance, hearing, and sight.

During the third, or Silurian, period of the Paleozoic, 425 million years ago, some animal forms, possibly descendants of the trilobites, but with much more flexible armor and highly developed eyes landed on the shore and survived.

They had no enemies and they were independent of the hitherto life-embracing oceans. Soon the coastal strips were to be alive with a multitude of creatures responding to the opportunities offered by the new environment. This invasion of the land by Silurian invertebrates was made possible by an advance party of plants that had firmly established itself ashore 70 million years previously. By developing rudimentary roots and by reproducing without the support of the ocean, the plants began to cover the rocks of the shoreline and banks of the lakes and rivers with slimy fronds while others evolved into low-growing, compact mosses. Although land plants had been slow to develop initially, by the opening of the Silurian period they entered a stage of rapid colonization and evolution. Another relationship between life, the lithosphere, and the hydrosphere was in the making. The earth had at last begun to look green.

The earth's long summer continued throughout the Silurian; the corals built reefs right up to the poles. Changes were afoot, however. Toward the end of the Silurian, the seas that had covered much of the land retreated, leaving many of the shallow-water environments high and dry. Some of the animals that had evolved in coastal seas and salt estuaries adapted to the new conditions, but many groups perished and became extinct—

In Carboniferous times lush vegetation died, decomposed, and became compacted into peat, then coal. Here, two modern scenes recall stages in that process. Above: part of a Brazilian rain forest. Right: section cut through peat in Galway, Ireland.

another example of the inexorable law of the survival of the fittest and most adaptable. The trilobites were beginning their trail downward into oblivion, but the corals, brachiopods, and crinoids continued to flourish. A nightmarish group of animals called sea scorpions, although they most probably lived in fresh or brackish water, reached their peak as the Silurian came to an end. They were large-headed, armored, and pincered, and ranged in length from a few inches to monsters over nine feet.

A gentle lifting of the North American continent entrapped the Salina Sea, an inland sea that covered the region from what is now Wisconsin to New York. This great salt lake alternately flooded and dried out over a period of several million years before it disappeared altogether, leaving deposits of millions of tons of salt and gypsum from 300 to 600 feet in depth. The geological gentleness of Silurian times in North America was not typical of other parts of the world. In what is now Europe the huge Caledonian mountain range was pushed up over an area reaching from the west coast of Norway and across the North Sea to Britain. Similar mountain-building activities were going on in Germany, France, North Africa, and Siberia. These upheavals drained the land of the remaining waters, except in Asia where a large inland sea remained.

The Silurian period was followed by the Devonian. Now the long summer ended and glaciers crept down the mountainsides. The Caledonian Mountains rose even higher than the present-day Alps; they were Europe's greatest mountains and marked the area of collision between Europe and North America during the formation of the continent of Euramerica that had begun when the continents started to move back in the Ordovician.

The Devonian was the period when the primitive vertebrates first came into their own in the seas and eventually waddled ashore to begin the conquest of the continents.

The primitive armored, fishlike creatures that appeared in the Ordovician and continued to develop during the Silurian, but without becoming significant members of the fauna in either period, began a rapid diversification. By the mid-Devonian, they had developed into several groups of fish both in the seas and in fresh water. Among them were sharks, two groups of bony fish, and several groups of armor-plated fish that had a period of success and then disappeared at the end of the era. The name *bony fish* distinguishes these creatures from sharks and their relatives, whose skeletons are composed of cartilage. Although the sharks seem to have appeared first in fresh water, they soon became

As the Carboniferous period began, backboned animals were gaining toeholds on dry land. Among the pioneers was Ichthyostega, *a short-legged amphibian known from fossils found in Greenland. Bony structures (for instance leg bones and tail-fin rays) show* Ichthyostega *must have evolved from air-breathing fishes that had limblike fins.*

almost exclusively marine and were the dominant vertebrates in the seas for several million years.

The bony fish, expanding in freshwater environments, developed in two directions. Some of them specialized in gill breathing, whereas others developed both gills and lungs. Among the latter were a group of special interest to us, because they are our ancestors. Their fins were not the ray-stiffened webs found among the most common fish of today, but were supported by internal bones attached to the central skeleton, somewhat in the way that the legs of frogs and salamanders are attached. It was the descendants either of these lobe-finned fish, as they are called, or of a closely related line that were to give rise to the land animals.

During the Silurian, land plants had evolved into many groups, and the Devonian landscape was clothed in forests of tree ferns, and strange trees covered with leaflike scales arranged on the tree like tiles on a roof, and giant horsetails. These were forerunners of the great forests of the Carboniferous period that was to follow.

The great Ordovician Appalachians in North America had been eroded away, leaving deposits of sand in the sea. In the west of the continent the crust was beginning to lift, bringing up with it the sediments lying in the great geosyncline between the continent proper and Cascadia. Deep beds of red sandstone were laid down in Europe. The layers, alternating between wind-blown and water-borne sediments, indicate that the Devonian climate alternated between arid and wet periods.

By the end of the Devonian period, the vertebrates were coming ashore, but these new land animals had to face a new set of problems. In the oceans, lakes, and rivers animals are supported by the surrounding water, but on land backbones and legs have to support the animals' weight, and this necessitates a number of skeletal modifications, particularly in the vertebral column.

The coming of the land vertebrates was a major step in the evolution of life, for it changed the emphasis of life from aquatic to terrestrial. In any evolutionary step the transitional period is necessarily a long one during which advances are made in small increments of change. Many of the early land animals must have been at home both in the water and on the land, breathing easily with both gills and lungs and able to swim as well as walk. These early changes would have been necessary even if the animals were to remain aquatic, for the long, Devonian dry periods made it increasingly difficult for fish to survive. With the exception of the sharks, the Devonian fish apparently remained in fresh water, and as the lakes and rivers dried up, either completely or seasonally, the survivors were those that had evolved a way either of staying alive until the water returned or of moving into a new pool or river. Only those animals that could directly breathe air to utilize the abundant atmospheric oxygen stood a chance of survival. During the Devonian a number of fish had, however, developed lungs capable of using atmospheric oxygen, showing once again that climatic changes could nudge evolution along in a given direction. Some of the descendants of the Devonian fish entered the sea, and as time passed their once-useful lungs atrophied and today remain as the air bladders in bony fish.

Fossil remains of amphibians, the first animals to live successfully on both land and in water, have been found in the Devonian rocks of East Greenland. Although these animals had many structural modifications to fit them for a terrestrial life, they still retained many of the characteristics of fish.

In the Carboniferous period the climate changed again. There was now water in abundance. Large areas of the continents were covered in swamps and swampy forests, full of rushes, tree ferns, giant mosses, horsetails, scale trees, and the first seed-bearing ferns.

Although the general picture of the earth during the Carboniferous is one of humid forests, there were deserts in some highlands.

In some places the trees were beginning to look more like the tall woody plants we are familiar with. Insects swarmed over the land in great profusion. Everything seems to have been large. Cockroaches three to four inches in length were everywhere, living on the dying and decaying vegetation. Dragonflies—some with wingspans of six feet or more—droned through the green and torrid forests. This was a time of change when those giants were exploring the possibilities of the newly developing ecosystem.

The humid forested lands, abounding in insects, were ideal for the slow-moving amphibi-

Fossil skeleton of Stereosternum, *an early reptile that lived in southern Africa in the Permian period.* Stereosternum *belonged to the mesosaurs: slender, aquatic fish-eaters three feet long, probably derived from ancestral reptiles known as cotylosaurs.*

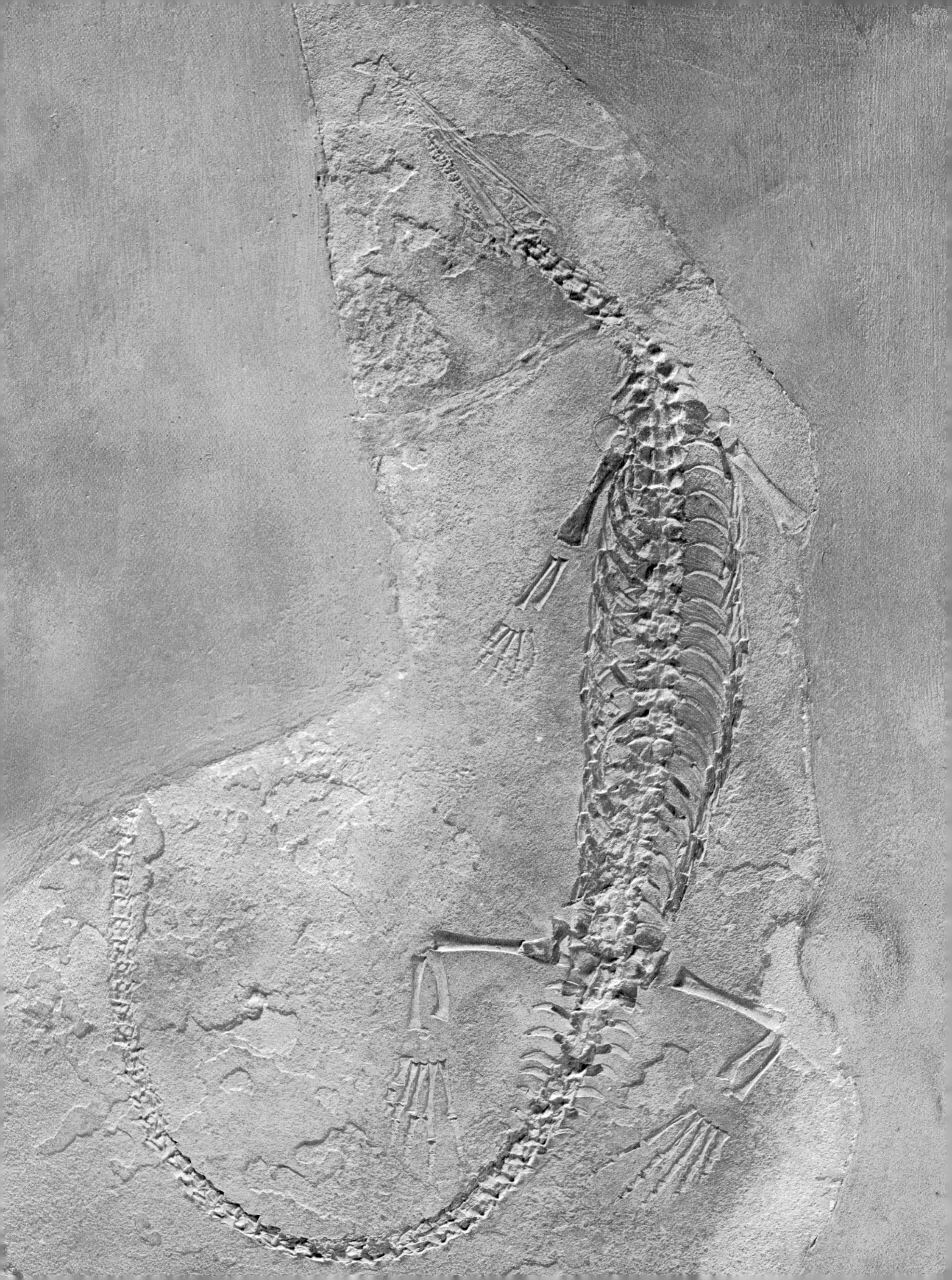

Fossilized Ceresiosaurus, *a Triassic reptile about three feet long that belonged to the nothosaurs. These reptiles had limbs that equipped them for both swimming and walking, and their digits were probably webbed. Nothosaurs belonged to a reptile group that eventually gave rise to the purely aquatic plesiosaurs.*

ans. Few competitors, the abundance of food, and an ideal climate provided the right conditions for their diversification, and many variants of the basic pattern appeared on the scene. The amphibians adapted to new niches as they appeared, but remained either wholly or semiaquatic. Some became even better adapted to an aquatic life. But alongside the amphibians lived another group that was adapting to a more terrestrial mode of life, the forebears of the reptiles.

By the Carboniferous period significant changes in reproduction had evolved. The eggs of the amphibians were, and mostly still are, fertilized externally, after being shed into the water. The early ancestors of the reptiles, however, had developed a way of protecting their young from environmental vicissitudes and to a certain extent from predators. The eggs were shell-covered and fertilized internally before they were laid. The reptilian egg provides the developing embryo with an adequate supply of food, and its permeable shell allows the occupant to breathe and to get rid of waste products. This innovation at last made it possible for vertebrates to take

up a completely terrestrial life.

During the Carboniferous, generation after generation of plants grew, flourished, and died in quick succession. Their remains lay in the swamps, where the process of decay slowly converted them into peat in which was stored the energy they had trapped during their lifetimes. The swamps, like the seas, were to become great stores of carbon, although not on the same scale or in the same way. As millennia passed, the huge peat deposits grew larger and were covered with sediments washed down from the highlands. Increasing pressure and heat from the overburden turned the peat first into lignite, and then into coal. The coal seams of today are the remains of generations of plants that lived and died in the great Carboniferous swamps. Throughout this period the earth had remained relatively quiet. A few mountain ranges had been lifted up, but without great crustal violence. Beneath the sediments, however, the heat in the mantle was building up and was once again moving outward to the crust. The earth began to shake as the layers of sedimentary rocks formed over millions of previous years cracked and buckled, shifted and folded. The Paleozoic era was coming to an end.

By the end of the last Paleozoic period, the Permian, all the continents were raised above the seas and the sedimentary rocks in the great geosynclines had been pushed up into great mountain chains. The Appalachians, Urals, and others had been squeezed into mountains as the continental sheets were pushed together. Weather patterns and wind currents changed by the newly risen mountains resulted in a colder, drier planet. Vast deserts presented life once again with a major crisis—adapt or become extinct. Many hitherto successful plants and animals could not adapt to the dramatically changing environment and made their exit from the stage of life.

The benevolent steady-state climate that had lasted for millennia now ended. By the end of the Permian, the surface of the earth was very different from what it had been for most of the Paleozoic era. Life struggled, adapted, evolved, and survived the tests. Many forms died out but new plants and animals evolved during the chaotic change. Among them were the reptiles, which had probably been saved by their leathery eggs. When the world dried out, most of the amphibians dried out with it as their eggs desiccated in the disappearing swamps.

As the climate became more arid and evaporation exceeded precipitation the inland seas grew increasingly salty and became dead seas. But before the drying out was completed, volcanic eruptions, heralding the movements in the mantle, burst out in what is now western North America. In the east the Appalachians continued to grow as thousands of feet of sediments, laid down in the offshore waters from the beginning of the Paleozoic era, buckled and folded into a new Appalachian range. These were very high mountains that stretched from Newfoundland south to Alabama. Volcanic activity in Europe spread vast lava flows over the continent. The sea invaded the low-lying regions, only to drain away again as the continent shifted and tilted. At the end of this activity the planet had but one large super-continent, Pangaea.

In the southern part of the super-continent, ice caps that had begun to accumulate in the closing stages of the Carboniferous period reached their peak of development, providing Australia with five ice ages. During the interglacials the sun drenched the land and lush forests grew and died, leaving their remains to form beds of coal buried under the southern island continent. In many parts of Australia volcanoes belched out their gases and ash to produce a landscape that must have resembled, on a larger scale, Iceland as it is today. In the South American and southern African regions of Pangaea, ice caps ground their way over the continent.

The plants that dominated the Carboniferous forests had almost died out, but the conifers and seed-bearing ferns lived on. The conifers became the dominant trees and were to continue down to our own times. The giant insects were gone, replaced by smaller species making their own niches in the new environment. The amphibians continued their decline but the reptiles were on the increase in numbers and diversity. By the end of the Permian they had almost achieved the dominance that they were to retain for 100 million years. Among them was a group of mammal-like reptiles which were to be the forerunners of the next dominant line.

The dominant species of the Paleozoic era either became extinct or were greatly reduced in number and the rise of new dominant groups of plants and animals to take over the vacant niches marked the beginning of the second era of the earth's history.

Two hundred and twenty five million years ago the Triassic period introduced the Mesozoic era and set the stage for the coming of the reptiles, mammals, and birds. The descendants of the reptilelike amphibians that had appeared during the Carboniferous period and survived the Permian chaos had grown numerous and diverse, and had begun to explore every ecological niche open to them.

One of the most important biological innovations during the reptilian advance was the appearance of plant-eating vertebrates, for hitherto the vertebrates had been carnivores. The fish shoveled for food in the sediments of river beds and bottoms of shallow seas, the amphibians and early reptiles snapped at cockroaches in the Carboniferous swamps, and the mammal-like reptiles ran up and down floating tree trunks to grab at fish and small amphibians. The small ancestors of the mammals continued to live by the rivers through the great Permian changes but as the climatic conditions altered in the Triassic they gradually changed from being predominantly fish-eaters to devourers of small land animals.

The trend away from dependence on water was a characteristic of the age that necessitated other changes in life style as well. Previously the energy flow in the ecosystems had been from plant to invertebrate and then to vertebrate. When the invertebrates first came ashore they lived on the intertidal plants, and thus were linked with the marine food chains. After the plants had covered the land, however, food chains gradually developed that were independent of aquatic environments. Plants were eaten by a variety of invertebrates that were eaten in turn by other invertebrates and these in their turn by the vertebrates.

The rise of reptilian herbivores heralded another stage in the development of terrestrial food chains, for from then on plants could be eaten directly by vertebrates and these in turn

Fossil Icarosaurus, *a Triassic gliding lizard named for the legendary Greek who fixed wings to his arms and flew. Projecting bones supported skin flaps that served as a parachute—extraordinary specialization for a lizard of such early date.*

would provide food for other vertebrates. The dependence on plants remained, but now the chains were much more diverse and although the animals and plants have changed, the system has remained to this day with little modification.

The Mesozoic era is often referred to as the age of the dinosaurs, but should really be called the age of the reptiles. Although the dinosaurs were the most dominant and spectacular of the reptiles, they were only one group of this amazing family. The evolution of the reptiles was rapid and within the geologically short spell of 40 or so million years they had produced thousands of different species. In South Africa alone over 1000 species of fossil reptiles have been found in a single rock formation, the majority belonging to a group of Permian vertebrates, the *cotylosaurs*, considered to be the first true reptiles. The cotylosaurs, which apparently had appeared with the early amphibians, perhaps in the Carboniferous but certainly by the Permian, retained many amphibian characteristics, a possible reason for their eventual extinction. Before they disappeared some of their offspring gave rise to the more successful lines of reptiles.

During the Triassic period the Tethys Sea flooded over many parts of the super-continent Pangaea, which now began to split into the large continental masses of Gondwanaland to the south and Laurasia to the north with the sea flooding into the split. During the 125-million-year Mesozoic era, the crust was subjected to earthquake shocks and eruptive volcanoes as the continental plates were tugged and pulled apart. By the final Mesozoic period, there were six continental masses. Antarctica had yet to separate from Australia, and India, which had been carved off Gondwanaland by plate movements, had to join with the Eurasian section of Laurasia before the modern shape of the world appeared.

At the beginning of the Triassic, southwestern Eurasia was covered by the Tethys Sea, which lay over a massive geosyncline that reached from southern Europe to the Himalayan region. The rest of Europe seems to have been arid, for the sediment deposits in the seas are composed of desert sands. In some areas, however, there are large deposits of salt and gypsum that suggest the presence of inland seas that dried up before the Triassic period ended.

The great southern continent of Gondwanaland was also geologically active. Belching volcanoes covered vast regions with lava flows and magma welled up into the sedimentary rocks. As a legacy of this period 30,000 square miles of sedimentary rock in Brazil are covered with lava flows in depths varying from 400 to 2000 feet. In the far south of Africa and in New Zealand glaciers covered the mountains.

On the far western edge of the North American area of Laurasia, there were a number of inland seas in which marine deposits were forming, while in other western areas winds eroded the rocks and covered vast areas with deserts of sand. In the eastern part of the North American section of the continent, the Appalachian mountains were being eroded. Streams and rivers carried the sediments down into the sea from which they would later rise as red sandstone characteristic of the Triassic period. Here too, huge lava flows spread out over the land as volcanoes released the pent-up forces in the mantle.

In this violent world the plants held on. Some just managed to survive, but many species perished and were replaced by new forms. The Triassic forests were composed of great conifer trees resembling the yews of today, and of cycads and ginkgo trees, species better suited to survive in the relatively dry climates of the period. In Arizona there lived some species of huge conifers, with trunks over 120 feet long and 10 feet in diameter.

The Triassic sea, too, was a place of change. New species of coral, mollusks, and crustaceans appeared, but the most important of the new species were cephalopods characterized by coiled shells. Among them were the *ammonites*, with shells similar to the pearly nautilus, and the *belemnites*, direct ancestors of the squids that swim in 20th-century seas. This new radiation of aquatic species provided a rich food supply for some of the reptiles that went to sea, probably by way of the lakes and rivers where they had lived on the bony fish. When these fish entered the seas, the aquatic reptiles followed them.

In this new environment both the fish and the reptiles increased in variety, and the marine reptiles in particular flourished on the abundant food supply. The *ichthyosaurs*, for example, became completely adapted to the marine

Not wood but stone—these fossil logs are the petrified remains of trees that grew in Arizona in late Triassic times, fell, became buried, had their wood replaced by agate, then were bared when erosion removed overlying clay. Growth ring studies show that the trees had stood in moist, warm valleys.

existence, and by the end of the Triassic had developed a fishlike shape. They swam by means of powerful thrusts of the tail and had dorsal fins and flippers as stabilizers, just as fish and dolphins have today. It is possible that the ichthyosaurs evolved straight from amphibian ancestors and were never a terrestrial group. Like some of the sharks they retained their eggs within their bodies until they hatched, and gave birth to living young. There are fossils of ichthyosaurs giving birth and fossils of females have been discovered with embryos in them.

There were other groups of sea-dwelling fish-eating reptiles living at the time of the ichthyosaurs, but they probably came ashore to lay their eggs. From these evolved another group of marine reptiles, the long-necked *plesiosaurs* that inhabited the seas right through to the end of the era, while generally improving their form for an aquatic mode of life. By the end of the Mesozoic the *mosasaurs*, another marine group that appeared in the Cretaceous period, looked like the classical sea-serpents of ancient legends. Their bodies were greatly lengthened, they had long, deep tails, long necks, and paddlelike limbs. Their small heads bore teeth suited to catching cephalopods.

In the second Mesozoic period, the Jurassic,

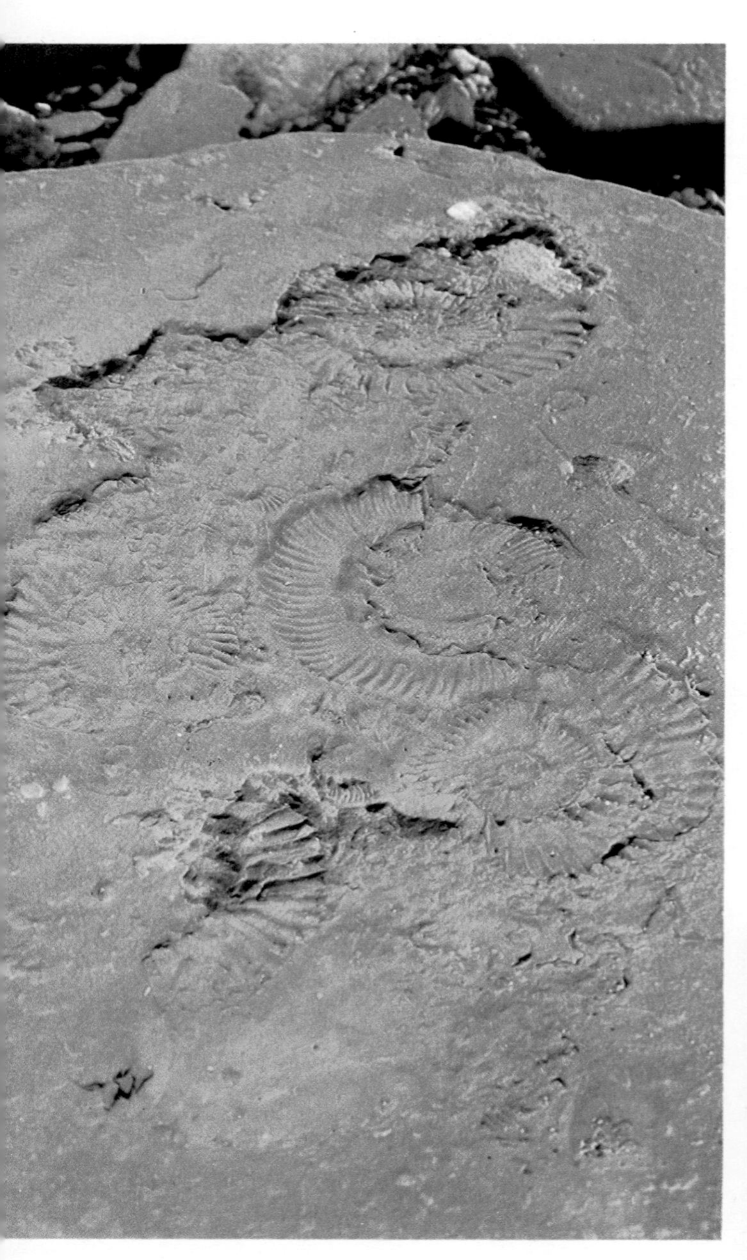

Above: ammonite impressions in rock at Kimmeridge, in southern England. Right: fossil ammonite, with a lustrous shell. The ammonites were marine mollusks equipped with a shell coiled in a flat spiral and divided into chambers, like that of the modern nautilus. They appeared as early as the Carboniferous period and reached a climax of development in the Triassic ocean, dying out entirely by the close of the Cretaceous period more than 60 million years ago. Ammonite shells ranged in size from dwarfs a mere half-inch in diameter up to giants six feet across. All told there were as many as 8000 species.

the typically arid Triassic climate moderated. Shallow seas once again spread over much of the land. Because large bodies of water moderate the fluctuation between extremes of hot and cold temperatures, most of Europe and North America, which were still attached, enjoyed subtropical climates. Corals once again extended their range, indicating that the waters were warm over a wide area.

In the shallow seas of Europe formation after formation of limestone and shale were deposited. North America extended much farther east than it does now. The lost eastern portion of the continent lies presently under the sea off New England. In the far west of North America the sea crept in through Canada and spread as far south as New Mexico and Arizona. As this sea withdrew later in the Jurassic, the rivers deposited their loads of mud, gravel, and sand in its emptying basins to produce a fossil-rich rock that includes thousands of dinosaur fossils.

Toward the end of the 50-million-year Jurassic period the earth rumbled and gave the western part of North America a violent shaking for the first time since the Precambrian in a series of mountain-building movements known as the Nevadian disturbances. These earth movements threw up the great sediments that had settled in the Cascadian geosyncline right through the Paleozoic and early Mesozoic eras into huge masses of folded sedimentary rocks. Much of this formation has been eroded, leaving the Pacific Coast ranges and the Sierra Nevada. The violence was so intense at times that many of the rock strata were turned over on edge. Huge flows of magma that welled up underneath the sedimentary rock distorted the crust and solidified into laccoliths of enormous size. Erosive forces have since worn away much of the sedimentary rock to uncover great granite domes such as those that stand out so imposingly in the Yosemite Park of California. As the magma and its mixture of rocks solidified, rich veins of metal ores formed along the cracks and fissures in the sediments. One of the most famous gold-bearing veins in the world, the Mother Lode that runs for nearly 130 miles in northern California, was formed at this time.

The Tethys Sea that covered southern Europe in the Triassic began to flood the greater part of Europe and reached as far west as Britain. One of the results of this marine invasion into Europe was the formation of a very important and

Fossilized shell belonging to a belemnite, from Jurassic rocks in central England. The shell has been opened up to reveal its chambered interior. Such shells were internal, and entirely covered by flesh. The soft body possessed six arms, which bore horny hooks. Belemnites were marine mollusks that belonged to the cephalopods and they were clearly close relatives of the squids, a type of cephalopod that has outlasted them. Belemnites were common in Triassic times, and reached their peak in the following Jurassic period when certain kinds were up to six feet long. But all had vanished by the end of the Mesozoic era.

Excavating a fossil plesiosaur, a large aquatic reptile that flourished during the Jurassic period. Bones grouped as in hands and feet represent the ends of limbs developed into flippers that propelled this long-necked, sharp-toothed eater of fish.

interesting Jurassic rock, the lithographic limestone that takes its name from a printing process, lithography, in which it was formerly used. This remarkably fine-grained limestone marks the position of the lagoons that had formed at the edge of the Tethys Sea. It is composed of marine sediments that were probably deposited in shallow waters under very quiet conditions. Many of the fossils found in the lithographic limestone beds in eastern Europe are in perfect condition, the most delicate of impressions showing entire creatures including their soft parts. Among them are fossils of *archaeopteryx*—the first feathered animal—crabs, turtles, crocodiles, and many marine reptiles, several species of flying reptile, and one lone dinosaur that appears to have waded out to feed in the shallow water and become trapped.

The Jurassic landscape would have had a certain familiarity to us. The forests were of pine and other conifers. There was, however, no grass. Instead, areas that would be grasslands today were covered with ferns and cycads. However, the plants of our own time were on the way; fossils of water lilies and magnolias have been found in the Jurassic rocks of Scotland. This was the first appearance of flowering plants bearing seeds enclosed in a protective case, which were to play such an important part in the rise of the mammals.

The Jurassic was the high noon of the dinosaurs. Some of them had by this time evolved into the largest creatures ever to live on land, with a bulk not to be exceeded by any animal until the great whales appeared millions of years later. The longest of them all was the herbivorous, 80-foot-long *diplodoccus*. Until recently it was

Mosasaur fossil from southeastern Brazil. Mosasaurs were flippered sea lizards of the Cretaceous period. Each had a long neck and body, and long, propulsive tail. The creatures possessed scaly skins and "elastic" jaws like those of snakes. Some of them attained a length of about 35 feet, and, as a group, the mosasaurs must have ranked among the fiercest of ocean predators.

Half Dome and other granite masses dominating California's Yosemite National Park originated as upwellings of molten rock forced up beneath sedimentary surface layers that have since been worn away. Half Dome arose during the Nevadian disturbance, which built the Sierra Nevada in late Jurassic–early Cretaceous times. Much later the peak lost a flank to attacks from glacier ice.

thought that this huge creature was confined to lakes and swamps where the water could support its great mass, but it now seems that they were able to carry their weight on the land. Another dinosaur, not quite so long but even heavier, was the *brontosaurus*, which, at a weight of 50 tons, was the heaviest land animal that has ever existed. Despite their bulk these giants and the smaller herbivores as well were preyed upon by the carnivores. One of the flesh-eating dinosaurs, *allosaurus*, a 30-foot, 14-ton giant, walked on its hind feet each equipped with three ugly-clawed toes. These and its powerful jaws with their three-inch teeth more than made up for its almost useless forelimbs. To sustain their bulk, these animals must have constantly hunted the herbivores. In response to predation the reptiles that had the capacity to develop protective armor were given an evolutionary push and gave rise to several groups of armored plant-eaters. Among those that evolved in the Jurassic was *stegosaurus*. It had a tiny head and huge body, 30 feet or more long, with two rows of loose bony plates along the top of its back and spikes on its tail. The function of the plates is not quite clear because it had them only along the top of its back, and the rest of its body was quite vulnerable.

Not all dinosaurs were large and not all reptiles were dinosaurs. Throughout the Jurassic period, reptiles both large and small took advantage of every environmental opening to spread and diversify. They filled all available niches and they even took to the air. Some of the flying reptiles were small but others, the giant *pteranodons* for example, had wingspans of over 50 feet. It is believed that the latter were not real fliers, but gliders that roosted on sea cliffs from which they could, by utilizing the air currents over the waves, glide over the surface just as the albatross does. The flying reptiles were not ancestors of the birds, which had already developed from another reptilian line, but a parallel evolutionary development to exploit the same niche.

Nature always guards against failure. There is always a new species coming up alongside the most successful design and waiting to fill its niche. Before the Triassic the only marine vertebrates were sharks. The marine food chains began with the microscopic plants eaten by the larger invertebrate animals and at the top of the marine food pyramid were the cephalopods, the most numerous and dominant species, despite the success of the sharks.

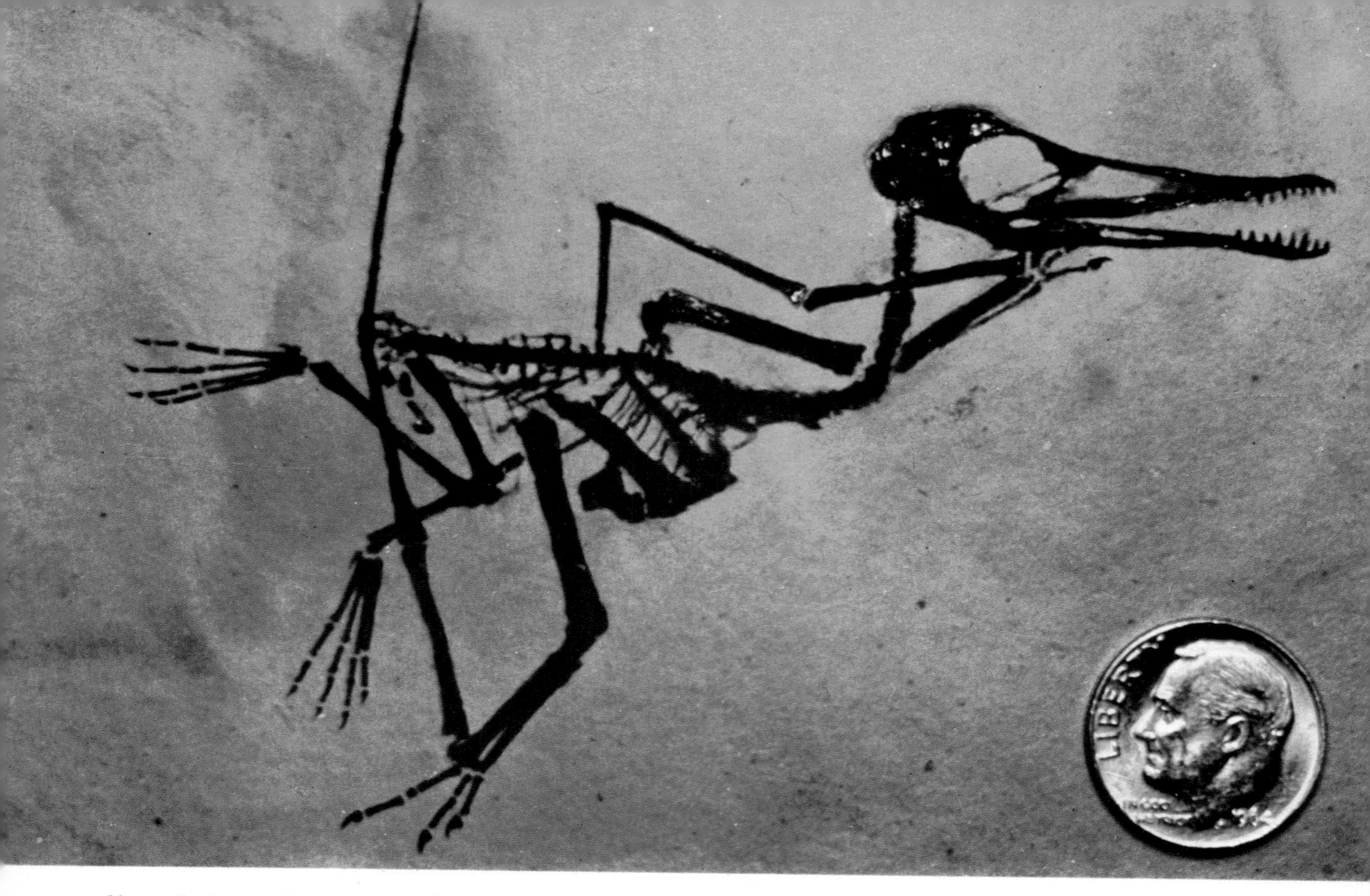

Above: the bones of the elongated fourth finger on each "hand" formed the leading edges for the skin wings of this pterodactyl (named from the Greek for "wing" and "finger"). Pterodactyls were small flying animals that evolved in the Jurassic period.

Below: Leptolepis, *a nine-inch-long Jurassic fish. It had jaws, a backbone completely made of bone, and other features that help us to put it among the earliest of the teleosts or bony fishes, a group to which most kinds of living fishes belong.*

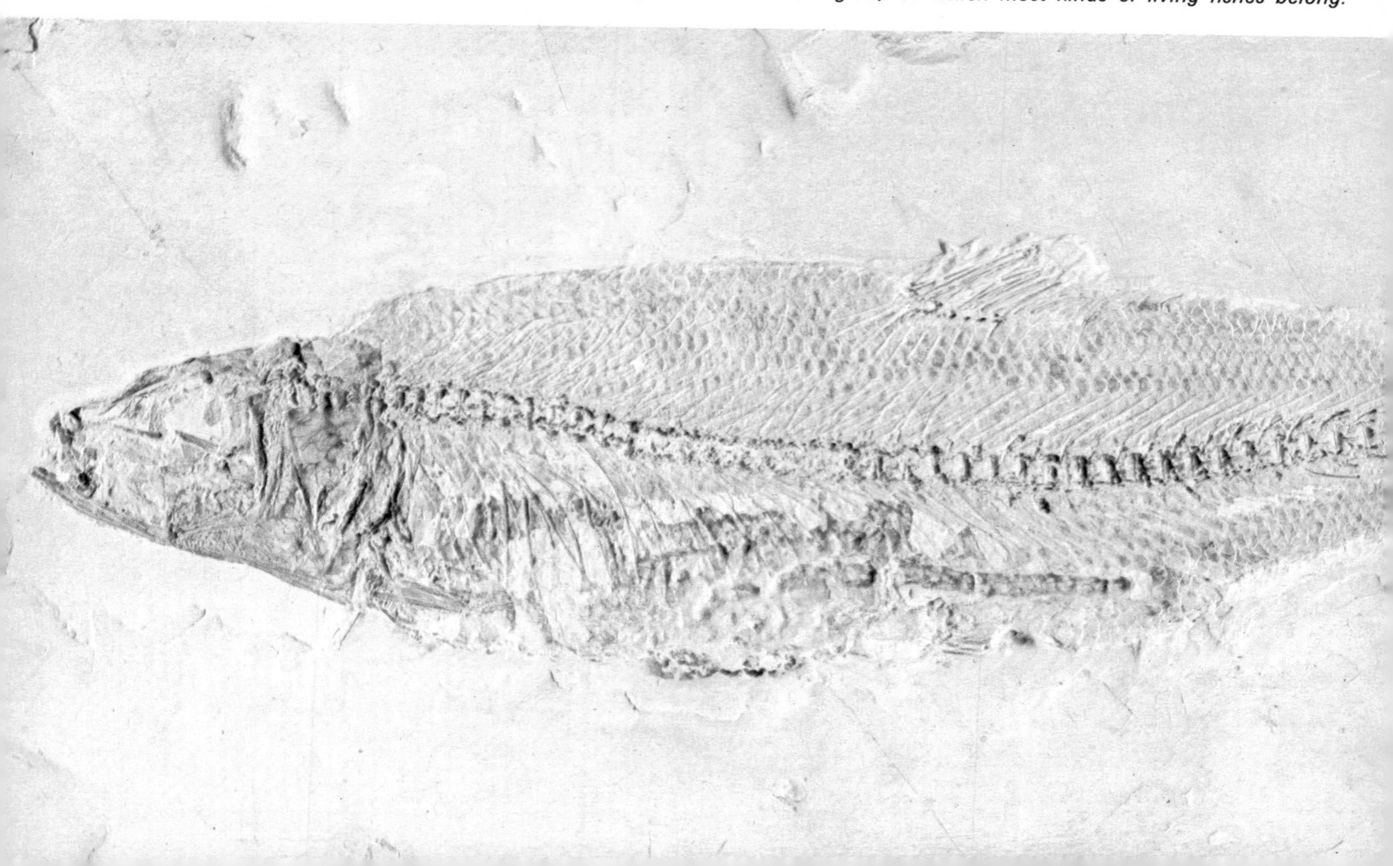

In the Triassic period, however, the freshwater bony fish invaded the seas where they were to soon rival the cephalopods. Close behind came the marine reptiles, and it did not take long for them to dominate the sea's food pyramid. The bony fishes were to diversify even more in the sea than they had done in the fresh water. Some became shellfish-eaters, others ate plant and animal plankton, yet others scavenged on the remains and debris left by the active feeders. At the top there were fish that preyed on other fish, so forming very complex food chains.

During the Triassic and Jurassic, successions of land plants and animals came and went. Among the plants, however, the first real seed-bearers were appearing. They were to become established as the climate became cooler and drier and the little proto-mammals that had also evolved during the early Mesozoic were beginning to specialize on the new plants, moving from clump to clump, and hiding from the carnivorous reptiles. These were the new ecological niches developing alongside the old ones that were now being worked out.

The Cretaceous period, which ended the Mesozoic era, marked the extinction of all the major reptiles in a comparatively short time—one of the most dramatic biological events ever to have happened. Why the dinosaurs died out so quickly is not at all clear. The spread of the seed-bearing plants must, however, have had some effect, because they have coarser stems and leaves than those on which the dinosaurs fed. The reptiles may not have been able to adapt to the new, harder plant materials.

Dominant animals at the top of any food chain are frequently the most specialized and consequently the most vulnerable to changing conditions. The dinosaurs were well adapted to the old mode of life, but not adaptable to the new. Only those reptile groups that could adapt survived. Among them were the crocodiles, and land and sea turtles, which have remained relatively unchanged since the Triassic period. Thus, as the dominance of the reptiles came to an end so the mammals filled the vacant niches and began to make evolutionary headway and the new flowering plants replaced their predecessors. In the sea, the bony fishes took over as the reptiles declined.

Whatever caused it, the great Cretaceous ecocatastrophy—the extinction of the dominant reptilian lines and the changes in vegetation—marked the end of a 125-million-year segment of the earth's story.

Seventy million years ago a new era, the Cenozoic (or recent life) era, started. We do not know when it will end.

The mammals that had begun to evolve from reptilian ancestors more than 100 million years earlier came into their own at the end of the

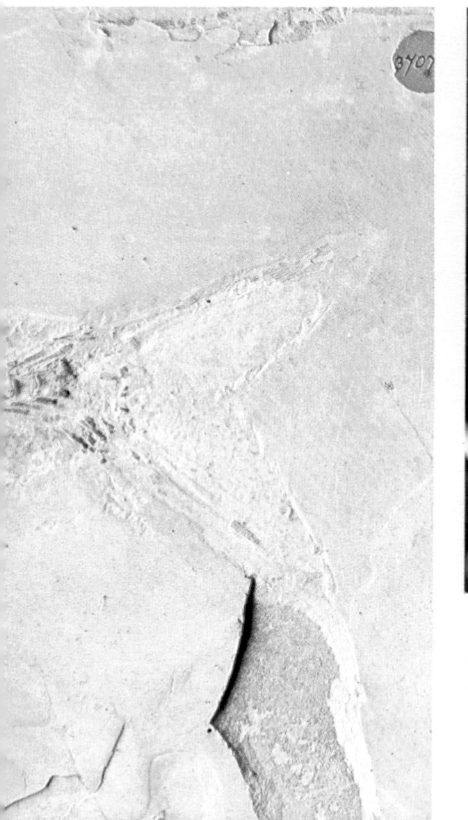

Flowers of the magnolia, one of the first known types of flowering plant. Appearing during Jurassic times, such plants have seeds enclosed in a protective case, which gives them a unique advantage. Flowering plants became the most successful of all.

The Cretaceous Period

1	Pteranodon	3	Monoclonius	5	Trachodon
2	Tyrannosaurus	4	Elasmosaurus	6	Struthiomimus

Nightmare beasts of Cretaceous days. Pteranodon *belonged to one of the groups of pterosaurs or "flying reptiles," now known to have had warm blood and fur.* Elasmosaurus *was a plesiosaur; 76 bones supported its snakelike neck. Of the four dinosaurs shown here,* Tyrannosaurus *(one of the largest-ever carnivores) is attacking* Monoclonius, *a herbivore inadequately protected by its heavily armored head;* Trachodon *was a duck-billed plant-eater;* Struthiomimus *resembled an ostrich, and it may have preyed upon the eggs of other dinosaurs.*

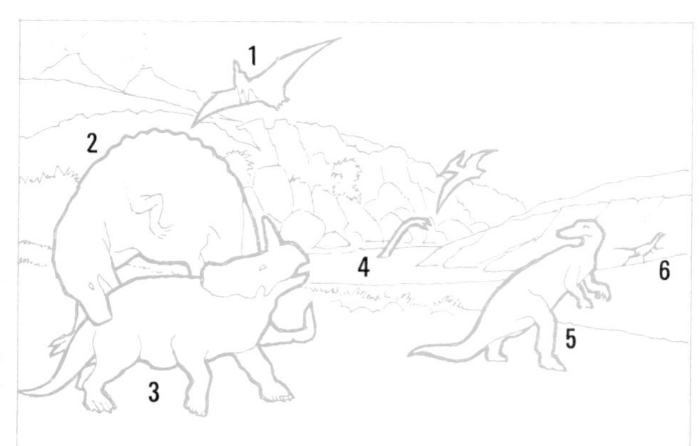

Above: fossil boa about three feet long from North America. Below: living crocodiles in the Murchison Falls National Park, Uganda. Unlike many groups of reptiles, snakes and crocodiles survived beyond the end of the Cretaceous period. Snakes evidently evolved from lizards in Cretaceous times. Crocodiles date from Triassic times and remain much as they were 100 million years ago.

Cretaceous. They had improved on the reptiles' form of reproduction, for instead of laying their eggs the females kept them within their body where they were fertilized and developed, free from interference and predatory pressures, except for those that were exerted against the female. Two major variations on this theme developed in parallel, the placental mammals and the marsupial mammals. Both fed their offspring with milk secreted by the mother, and evolved complex family relationships that ensured the care and well-being of the young during their most vulnerable period.

The mammals' covering of fur was a particular asset at the beginning of the Cenozoic era for during its first two epochs, the Paleocene and the Eocene, the world grew cooler, foreshadowing the ice ages to come. The mammals' warm blood, combined with insulation to help maintain body temperature at a constant level, gave them a distinct advantage in a world that was to grow cold. The birds, too, were warm-blooded and also had an insulating cover. Feathers were at first a form of insulation and then were later adapted for use in flight. Although birds ruled the air, some mammals also evolved as flyers with reasonable success. Egg-laying was retained by the birds, a characteristic that made them more vulnerable to predators than the mammals.

At the start of the present geological era the globe looked much as it does today. Great mountain-building movements continued to rock the continents. As the era progressed, the Alps and the Himalaya buckled up as the crustal plates crashed together. The Andes and the Rockies grew in stature as the land lifted. Volcanoes poured out lava and in North America lava flows covered 200,000 square miles of land between the Pacific Coast ranges and the Rockies to a depth of thousands of feet. The Appalachian Mountains

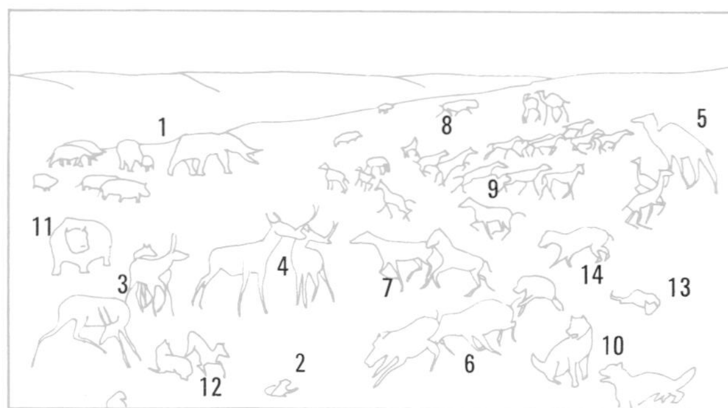

The Pliocene Epoch

1 Amebelodon
2 Epigaulus
3 Craniocerus
4 Synthetoceras
5 Megatylopus
6 Prosthenops
7 Neohipparion
8 Aphelops
9 Pliohippus
10 Osteoborus
11 Teleoceras
12 Merycodus
13 Pseudaelurus
14 Hemicyon

This imaginary Pliocene landscape in North America shows now-extinct mammals that were close kin of living forms: (1) mastodons; (2) a rodent; (3) and (4) deerlike animals; (5) camels; (6) swine; (7) and (9) horses; (8) and (11) rhinoceroses; (10) dogs; (12) prongbucks; (13) a cat; (14) a "bear-dog."

had been eroded away, but the present Appalachians were lifted up and the erosive forces started sculpting the now familiar landscape.

The climate gave way to cooler conditions, and temperatures continued to drop until it became so cold that glaciers formed in the higher latitudes. About a million years ago the Pleistocene epoch arrived, driving the mammals south and generally squeezing the biosphere into a narrowing band, though there were to be warm periods again during which it would expand. The major groups of plants that had appeared in the Cretaceous flourished and extended their range, limited only by the changing climate.

Two new developments in plants have occurred in the present era that were to play an important role in the subsequent history of the biosphere and crustal rocks. The first appeared in the sea during the Eocene, 58 million years ago. It was a tiny plant form called a *diatom*, a single cell of photosynthesizing protoplasm in a beautiful shell of silica. When these plants die their shells fall to the sea floor and accumulate, and during the past few million years they have formed thick deposits of a light, porous rock called *diatomite*. Although the diatoms have been and still are important rock builders, their function in the continuing history of the planet goes beyond rock building. These tiny plants are the grass of the seas, the energy base of marine food chains, and they are an important source of the recycled oxygen upon which all life, both aquatic and terrestrial, depends. The second plant development appeared later, during the Miocene, some 25 million years ago. It was the arrival of *grass*, heralding a new organism capturing solar energy and preparing nutrients at the foot of the terrestrial food pyramids. The grasses were to be a very successful line and soon evolved into a number of different forms to occupy a variety of conditions. The mammals responded by producing the grazers.

The road along which life evolved has been a long and bumpy one; it took over 2000 million years to make the first cells, another 1000 million for multicelled life, and a further 600 million to reach the complexity and rich variety of life and living associations in the world today. Born of the earth's crust and sustained by it, life has played an important part in the distribution and cycling of the crustal elements. The world today contains a complex mixture of carbon compounds in a continuous state of synthesis, transformation, and decomposition, controlled and maintained by the phytoplankton in the sea and plants on the land that capture the energy of sunlight to manufacture complicated organic molecules from carbon dioxide, water, and mineral nutrients. The processes that provide for the circulation of elements through living organisms are the outermost cogs in the biogeochemical machine that involves the entire planet. Yet, only a minute amount of the earth's chemical elements is actively circulated in the biosphere. The bulk is locked in the deposits that have taken millions of years to accumulate. The cycles circulating through living systems are coupled by a very low gear to the grander geological processes; the rate of replacement of calcium, for example, is measured in hundreds of millions of years. Life has modified the earth's cyclic scheme to enhance its own potential. From an early random distribution, the elements have now been so concentrated that they are recycled through the geological processes to be used again in future time by future forms.

Moorland grasses on Exmoor in southwest England. The appearance of grasslands in Miocene times favored the rise of grazing mammals—mammals with high-crowned cheek teeth designed to withstand wear produced by chewing food with a high silica content.

The Human Odyssey

In the forests of the last days of the Cretaceous, some 70 million years ago, a group of small insect-eating mammals was developing into the line that would one day become the dominant animal group. These mammals, the early primates and our remote ancestors, had problems, for their ecological niche was threatened by fierce competition from the rodents that were evolving at the same time. They did manage to hold on, however, until, about 36 million years ago, there appeared a group of primates that were to become the direct ancestors of man.

Of the multitude of plants and animals that have lived on the earth, man is the only single species to have risen to an almost unassailable position as the dominant animal. Prior to man, dominance had been attained by groups of related species. The dinosaurs, for example, were an assemblage of many different species of reptiles, some as different from one another as man is from the dog.

Man's earliest recognizable ancestors took to the trees, which was to prove advantageous because it provided time for those little mammals that had not yet advanced to the monkey stage of development to acquire a number of useful attributes. To move through the trees, they had to resort to jumping with accuracy from one branch to another, which made it a decided advantage to have two eyes able to focus on the same branch. By a process of elimination—literally dropping the unsuccessful—the primate ancestors whose eyes were closer together survived. Gradually, over a period of several million years, their eyes moved from the side of the head toward the front, providing binocular vision. The improvement in sight decreased the importance of the sense of smell; long noses or snouts were no longer necessary and faces became flatter. With binocular vision came improved muscle coordination and, important in an arboreal existence, a sense of balance. A clawed inflexible foot is of little use for branch-to-branch travel. Thus claws went out and flat nails came in. Above all, the ability to grasp branches was essential and as a result the opposable thumb evolved and improved. In time this adaptation also provided the means to grasp and handle other objects—food, tools, plows, swords, gear levers, and pens.

Sometime during the period that marked the end of the Miocene and the beginning of the Pliocene period some of the primates came down from the trees and took up life on open savannas in Africa, Asia, and western Europe. One type in East Africa and India was showing signs of the man-primate to come. This protohuman ran about on its hind limbs. Although its teeth were next to useless as defensive or offensive weapons, it used its grasping hands to lift and use sticks and stones for defense and for catching its prey. This was possible because its forelimbs had been freed from their role of supporting the body.

In the struggle to survive, certain characteristics already present in the protohumans came to the fore: his hands became more flexible, his legs became stronger, and his brain grew larger. With his sharpened wits he observed the migration of animals and where the predatory bears and cats rested. He discovered better fruits, nuts, and roots, and learned how to use sticks and discarded bones as tools to dig roots and grubs and to kill other animals. All this took another few million years.

Then at some point, perhaps half a million years ago, some ancient man held in his hands a burning stick. He could carry it and spread the fire. He observed that after fires had passed, new grass grew and many fruit-bearing plants appeared, and had not the other animals fled the fire just as he had done? The use and control of fire was the first and most fundamental discovery made by mankind. With it men could cook meat and plants, thus widening and tapping a greater reservoir of energy and nutrients.

A quarter of a million years ago several forms of man were living, and although they were similar in looks and posture to those of us living today, most were to be eliminated either by modern man or by the changing climate of the ice age. The successful species, *Homo sapiens*, adapted to the new conditions. They moved south ahead of the advancing ice and survived to follow it back as it retreated.

Early modern man formed societies to ensure that his kind kept together, for as an individual

Aboriginal rock paintings in Cape York, northeast Australia. By depicting figures in wished-for situations primitive man tried to use magic to mold his environment. At this stage man made relatively little mark on the face of the earth. But as he learned new ways of exploiting nature's resources his impact increased. Today, man's activities affect most parts of the earth's crust.

he was still a very inadequate hunter and a cooperative group was needed to round up animals for the kill. Life in a community also provided protection and security for the young, who were particularly vulnerable during the first few years of their lives.

Gradually man's awareness and imagination increased. He thought he saw images in the glowing embers of a cooking fire, and trees seemed to change into strange shapes in the evening twilight. He felt the rumbles as the earth moved and heard strange sounds coming from the hills, forests, and heavens. He invented elaborate ceremonies, used colored clays and fragments scraped from rocks to paint pictures of his prey, and marked on them where he hoped his weapon would strike the fatal blow. He painted and made models of the female form, hoping to encourage her to be fertile. Above all, wanting to be master of his house, he sought power and guidance in superstition and religion. As man grew in competence and lived increasingly in groups, he needed to communicate ideas to his comrades. Once speech was established it speeded up his intellectual, technological, and social evolution.

Fires lit by man destroyed the vegetation that had evolved and established successful biological regimes. Trees and shrubs fell under fire and ax just as their forerunners had fallen under the impact of the changing climates. Now, because of man, there was a difference in the rate of their disappearance—it was measured in decades and centuries rather than millions of years. The shifting continents and changes in the physical world were no longer the only cause of extinction of unadapting species. Man had now become the nemesis of other living things.

Man was to disrupt for all time the ancient biological cycles that had slowed up the process of weathering and erosion. While he had been a hunter-gatherer and his numbers were limited by the supply of roots, fruits, and the animals he

Left: goats—said to eat almost anything—are helping to strip vegetation from this barren farmland in Upper Volta. Problems caused by the overgrazing of man's flocks and herds are especially acute in Saharan-fringe nations, where periodic droughts tend to curb plant growth, and intensify the grazing and browsing pressures on what plants there are. Stripped of their leaves by goats and cattle, shrubs and low trees soon die, and soil held in place by their roots drifts off on the wind.

Below: Sahara Desert Sand ridges that resemble waves. Like the waves of the ocean, desert sands may move with the wind, particularly where there are no strongly rooted plants to hinder their progress. Thus, where overgrazing and drought have killed plants at the edge of the desert, Saharan sands have blown south to engulf large tracts of land.

A derelict Oklahoma farmhouse recalls the Dust Bowl disasters of the 1930s. Overcropping followed by drought provoked soil erosion that turned productive fields into dusty deserts and drove out thousands of farmers. Some people never came back.

could catch, man's effect on the environment was no more than that of any other predatory animal. It is a truism of nature that predators do not exterminate their prey, and man at that time was no exception. He was part of a balanced system meshed with all the other cyclic systems that had evolved over the millennia. Soon, however, he stood alone and burned the vegetation.

Over the vast areas of the world, winds followed the fires and carried away the soils that had been protected by plants for thousands of years. Vegetation conserves the soil and keeps erosion in abeyance by holding the soil in place, thus protecting the bedrock beneath from the ravages of air and water. When the covering vegetation is removed, the humus in the soil dries out and the complex soil structure is broken up, disrupting the intricate relationship of the soil organisms. In time, as the long-conserved moisture evaporates and the dry soil blows away, biologically nonproductive deserts are formed.

The effects of the fires of the hunters and primitive agriculturalists were small compared with those of the shepherds. Man domesticated goats in the Middle East around 7000 or 8000 years ago. By cutting and burning the forests to increase the grasslands, the shepherds of the Middle East and Mediterranean lands were responsible for destroying vast forests. At the time of Hannibal's march against Rome during the Second Punic War, much of the North African region was covered in tracts of forest, roamed by elephants and lions. The climate of the Mediterranean region has hardly altered during the past 6000 years, and it cannot be claimed to be a main factor in the changes the area has undergone.

China's Yellow River takes its name from its immense load of yellow sediments, derived from the loess ("yellow earth") regions through which it passes. Deforestation and agriculture have made these regions especially liable to river erosion.

An industrial scene at Ebbw Vale in south Wales, one of the coal-rich regions of the British Isles, where the Industrial Revolution began. As factory production spread, it created mass markets for manufactured goods and helped transform small, scattered, rural populations into much larger and more strongly urbanized communities.

A New York City sidewalk in the rush hour helps us to realize how mankind's numbers have multiplied since the early Stone Age when he was only a rare foraging creature. Now, cities hold millions of people, thanks to modern mining, manufacturing processes, and, above all, modern methods of food production—activities that affect large parts of the surface of the globe and place great strains upon the crust's limited resources.

The goat is the most ancient of man's partners as an agent of environmental change, but overgrazing by sheep and cattle has played almost as important a part. The forest cover of Europe had disappeared almost entirely by the 18th century. In Britain many of the open moors are the remains of once lush forests, reduced to wet deserts by overgrazing and burning. During the 18th century the same catastrophic practices denuded Iceland of the natural plant cover that once clothed a large part of the island.

The greatest of all sand deserts, the Sahara, increases in size every year. Now even small changes in climatic conditions can lead to catastrophe—a drought of short duration in an area of reduced plant cover, such as savanna, can bring the cycling of nutrients and water to a complete stop. Because the plant cover has been impoverished by overgrazing and the drought has stayed a little longer than expected, the Sahara is advancing south and east by 10,000 acres a year. If only because of its speed, the most dramatic loss of soil cover in history was experienced in the early part of 1934. The cause of this disaster was the effect of an unexpectedly severe drought on the ruined soils of parts of the south central states and eastern plains region of the United States. Devoid of natural plant cover and crops

to hold them together, the soils were blown away on the prevailing westerlies. Through all that spring, summer, and fall, millions of acres were stripped of their precious soil.

The dusty soil was carried high into the atmosphere, darkening the skies over Iowa and Illinois, producing spectacular sunsets even as far east as Washington D.C. and New York City. The heavier fractions fell in the east of the country, smothering cattle and crops. This was the origin of what became known as the "Dust Bowl." The drought went on for some years, and families were forced to leave their farms and homes for good. Thousands of homeless people joined the ranks of migrants to California.

Even now, the same time-honored practices of burning and overgrazing are still being carried out, depleting the soil cover of the continents still further. In some parts of Africa the number of cattle a region can support without damage has been exceeded by 100 per cent or more. In addition to the cattle, there are also four times as many sheep and goats. In some regions of East and South Africa over 70 per cent of the soil surface is affected by overgrazing.

The introduction of settled agriculture 7000 years ago brought conflict between the new farmer and the older pastoralist. It also heralded an even greater conversion of forests into open spaces to make room for crops. Gradually agriculture provided the human communities with a surplus of food, which, as techniques improved, could be stored for later use, especially at times of crop failure. It also brought with it a new element in man's affairs: trade. Agriculture freed man from the time-consuming need to hunt and to gather fruit, and provided freedom to diversify his employment, producing craftsmen, soldiers, merchants, lawyers, artists, scientists, coach drivers, and politicians.

Since the advent of agriculture, the human population has increased enormously, for like those of all other organisms man's numbers are dependent on food supply. At different times the human population has overtaken the food supply. It has left, and still leaves, little margin for natural and localized fluctuations in environmental conditions. Consequently from time to time famine has brought a sharp check to progress. At times the diminishing resources have triggered off tribal and national rivalries, causing wars between the competing groups. As time has gone on, however, famine and war have become less effective as a means of controlling the numbers of human beings.

The degree of mastery over his environment that man achieved during the years immediately following his discovery of agriculture gave him a great advantage over the less well endowed species. In less than 10,000 years he has effectively changed the environment to suit his own needs. Only the insects, some microorganisms, and a small number of plants and vertebrates threaten man's position as the world's dominant animal, and even these are marginal.

It is not man's agricultural practices alone,

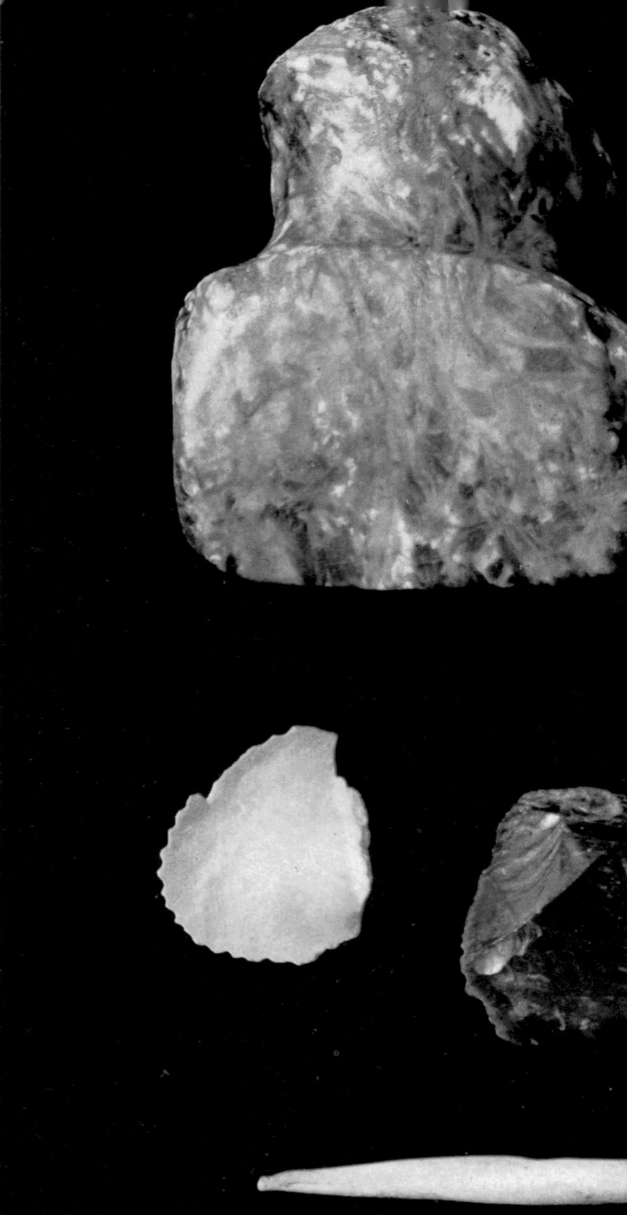

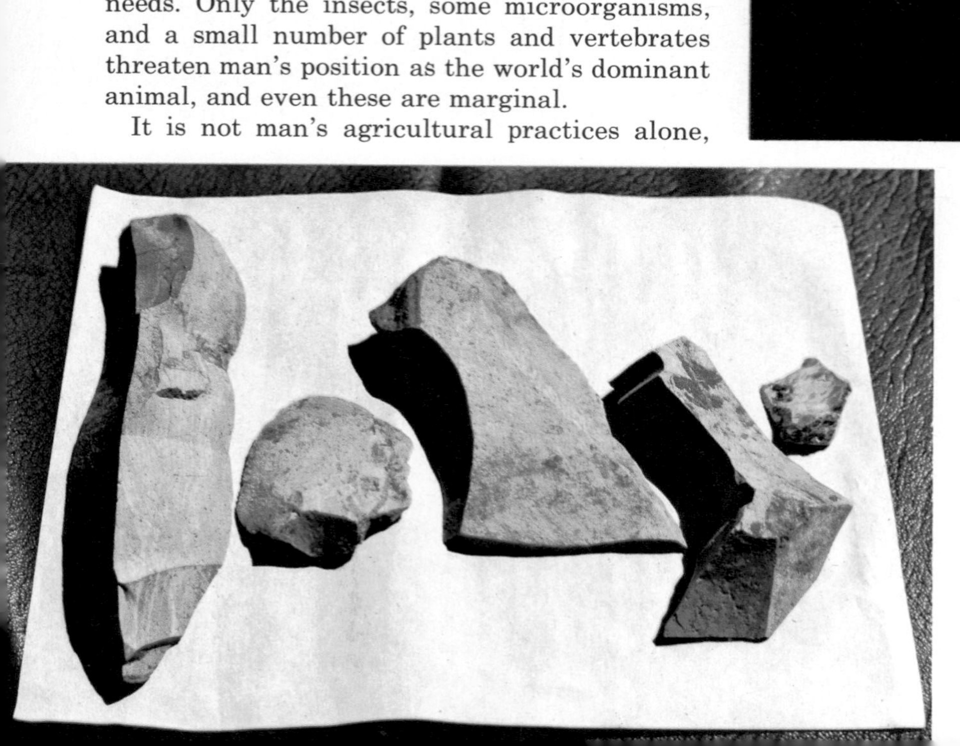

Man's exploitation of the raw materials of the earth's crust began when Stone Age people chipped surface pebbles into tools with predetermined shapes. Above: American Indian flake and bone tools from Arizona's Colorado Plateau region. They represent (top row) an ax and three clubs; (middle row) four scrapers and four awls. The bone instruments may have been used to fashion stone tools. Left: less finely worked stone implements made by Australian Aborigines.

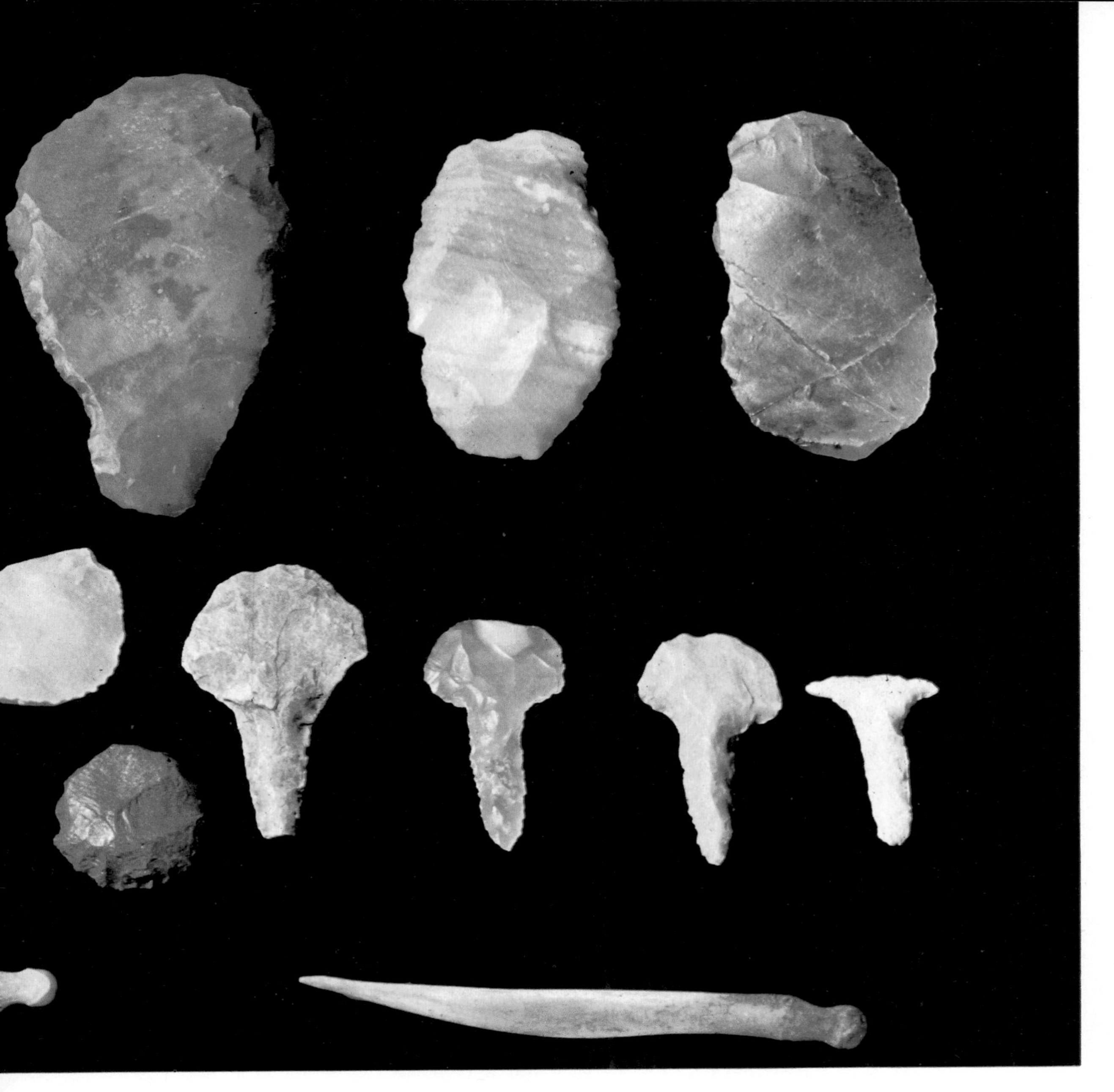

however, that have changed the nature of the surface cover. His need for fire gave rise to a need for fuel. The first abundant fuel resource was also vegetable: rushes, grasses, and wood from trees and shrubs. The discovery of metal working opened up a whole new way of life, however, and man required increasing supplies of fuel to smelt the metals.

In China, one of the ancient birthplaces of technology, there are large areas of poor soil cover close to the bedrock, as a direct result of the destruction of forests for industry and agriculture over 3000 years ago. The Chinese tried many times to reintroduce forests, but the demands of a growing human population have defeated all attempts.

Man has changed the biosphere over a large portion of the globe, for he has seized a massive proportion of the energy and nutrient supply for his own use. In so doing he has aided the weathering and erosive forces that are now reducing productivity, and is once again close to disaster. Except for the small percentage of food supplied by marine resources, an enormous population—over 4000 million—is totally dependent on an energy-consuming agricultural system. Widespread famine may once again impose a check on his numbers.

Not only does man depend on the earth's crust for his sustenance, but he has also built upon it a massive and complex industrial civilization. The wealth and security of his first cities—perhaps his greatest creation—were dependent upon agriculture. The first cities were sited in areas where he could settle and grow crops in the nutrient-rich river valleys and deltas of southwestern Asia: the Tigris and the Euphrates, the Jordan, and the Nile.

Food production by any means requires tools and weapons and this is where opposable thumbs came into their own. The hands that grasped the fiery stick could also hold digging tools and weapons made of wood, bone, and horn. Man's first tools were those he found, the odd stick or stone, but seeing what could be accomplished with tools of certain natural shapes, he reproduced them by crudely chipping the materials at hand instead of searching for the desired forms.

Examples of stone tools have been unearthed in the Olduvai Gorge in Tanzania dating back 2 million years. Peking man and his contemporaries made hand axes $1\frac{1}{2}$ million years later, but the real beauties of stone-tool craftsmanship were the finely chipped and polished blades and axheads made between 100,000 and 10,000 years ago by Old Stone Age toolmakers in Europe and Asia. They produced very small arrowheads and small knife-shaped points of flint, quartz, and obsidian as well as elaborate bone fish spears, fish hooks, and needles. Personal adornment also seems to have begun at this time, and necklaces made of the canine teeth of carnivores and other bones are found in graves of the period. Besides being skilled craftsmen, these people had a well-developed artistic sense; they carved elaborate designs on antlers and painted animal portraits on the walls of caves.

It may have been his interest in self-adornment that opened man's eyes to the potentially beautiful objects around him. Sitting by the side of a stream waiting for a fish to come within a spear's throw or just idling away a moment, one of our early ancestors saw a bright yellow pebble and plucked it off the stream bed. It did not have the feel of stone, but it was pretty. In such a way, one could conjecture, gold entered the lives of primitive men. They found that the soft metal could be worked into shapes and designs and very soon it became a much-sought-after material. Thus metal working was added to man's growing list of accomplishments.

Although the discovery of gold and its malleability was a limited step forward, it gave man the "feel" of metals. Given this knowledge he soon realized that copper was more useful than gold, if not as pretty. Copper may have been first discovered by accidental extraction of the metal from its ores in stones used around the cooking fires of Neolithic men, or it may have been the native metal that attracted man's atten-

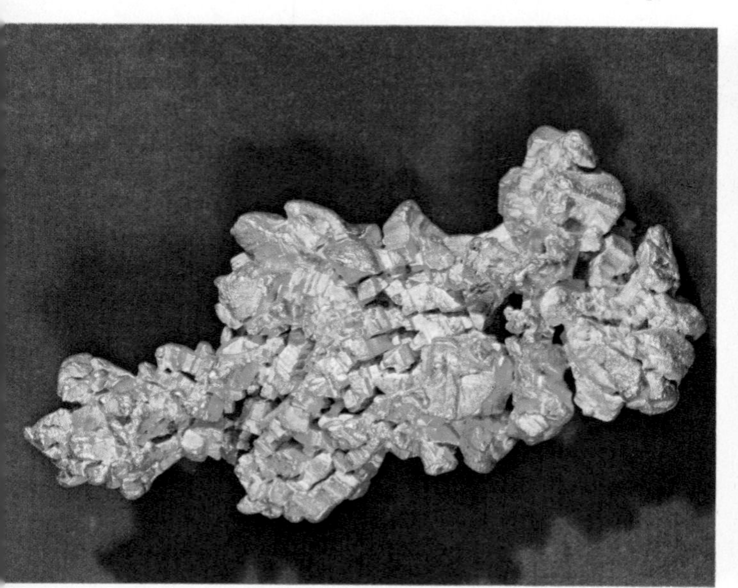

Gold—attractive, easily worked, untarnishable in a pure state—was probably the first metal to be prized by man. Left: gold nugget, the form in which Stone Age man may have first noticed gold. Right: beaten gold work from ancient Peru. Peruvian Indians also mined tin, copper, and silver. They learned to smelt the last two from their ores and to fashion metal ornaments by casting.

Copper was the first metal obtained by man entirely for practical purposes. Early copper workers used only small quantities of copper, where they found it in a fairly pure state. We still value copper, though not now for knives or ax-heads but for wires and pipes and a wide range of hardware. Copper exploitation on today's scale became possible through the invention of sophisticated techniques for efficiently extracting copper from its ores. Above: sprinklers at Santa Rita, New Mexico, wash four pounds of useful copper from every ton of copper waste. Right: molten copper being poured during a copper smelting process.

tion. (A native metal is one found in its metallic state, not as an ore that must be treated by heat or chemical means to release the metal.) Once the value of copper tools was realized, the search for its ores and for ways of getting the copper out of them were pursued with vigor. Somehow, probably in early copper extractions, molten copper was mixed with molten tin and produced bronze. Such a chance combination provided man with the first alloy. How similarly are some of the modern scientific discoveries made?

Bronze provided a hard metal with which to make and perfect weapons, tools, and pots. The Bronze Age announced the era of metal technology, which—in spite of plastics—is still the basis of modern industrial civilization. The refined techniques required for the extraction and alloying of copper and tin prepared man for the discovery of iron, and of ways to smelt and forge it into tools and weapons harder than bronze.

Soils, caves, stones, metals—only one major discovery is missing from the repertoire of man's utilization of the earth's crust, and that is the use of clay for pottery and building materials. The use of clay for making pots may have been discovered accidentally, perhaps when some mud stuck to one of the rush baskets that were the earliest of receptacles, coming into use just before the dawn of agriculture. When the sun-baked clay was removed, it retained the shape of the basket and the impression of the reeds with which it was made. It was a crude pot, but despite its clumsiness, the prototype for a new art and industry. The accidental reed designs were later copied by hand and formalized as a design style.

The center of man's dispersal was somewhere in the Middle East region, where the three old-world continents meet. There, too, his agriculture and primitive industries began. While Europe and North America were in the grip of the Pleistocene Ice Age, much of the Middle East was drenched in what must have seemed to be

New York, like other modern cities, is largely made from ingredients of the earth's crust reshaped or reconstituted to form masonry, bricks, concrete, glass, and steel. Such artificial environments may cover huge areas where level land abounds and stimulus to urban growth is high. But lack of space on New York's Manhattan Island called for vertical expansion, resulting in the world's best-known group of skyscrapers. Left: the skeleton of a new building dwarfed by the huge Rockefeller Center. Above: aerial view of skyscrapers by Manhattan's Battery Park.

Structures designed for our urbanized industrial environment modify the natural environment in a number of ways. Left: skyscrapers produce artificial canyons through which air is violently funneled, and a city as a whole heats the local atmosphere. Above: Graveley Interchange (aptly nicknamed "Spaghetti Junction") at Birmingham in England. Highways are among the city's arteries and veins, through which labor, food, and raw materials reach an urban area and manufactured products can be exported. To service today's city, engineers must thus bury immense tracts of land under strips of concrete or tarmacadam.

perpetual rain, and much of the Persian plateau was covered by an inland sea. When the glaciers retreated, the rains stopped and the land began to dry out. The inland sea evaporated, leaving a great desert. But for a long time before the sea disappeared, rich grasslands grew on its shores and along the banks of the rivers draining into it. The lush pastures attracted wild grazers and then hunters, both men and carnivores.

The men who settled on this plateau were skilled hunters. They could make crude pottery and build rough shelters in natural or man-made caves carved out of the mountainsides. They spun fabric, decorated themselves, and buried their dead. This settlement occurred about 10,000 years ago and it was here that domestication of plants and animals took place.

As the farming communities developed, so did housing. Houses built of hand-molded pancake-shaped bricks began to appear, and as the millennia passed, pottery and bricks were refined in both form and function. As an interesting sidelight on the development of the utilitarian clay pot, there has been speculation that the first wheel and axle may have been a potter's wheel. Metal-working skills also became more sophisticated and jewelry began to show foreign influence as shells and precious metals from the northeast of Persia came to the area through trade carried on by the newly risen merchant class. All these developments improved the quality of life beyond any imagination of men of the Old Stone Age. More important by far were the beginnings of written language. Symbols scratched into blocks of clay marked a major turning point between primitiveness and civilization.

Man was now established as monarch of the known world; all he had to do was to spread, colonize, and consolidate, which he did successfully throughout Africa, Asia, the Americas,

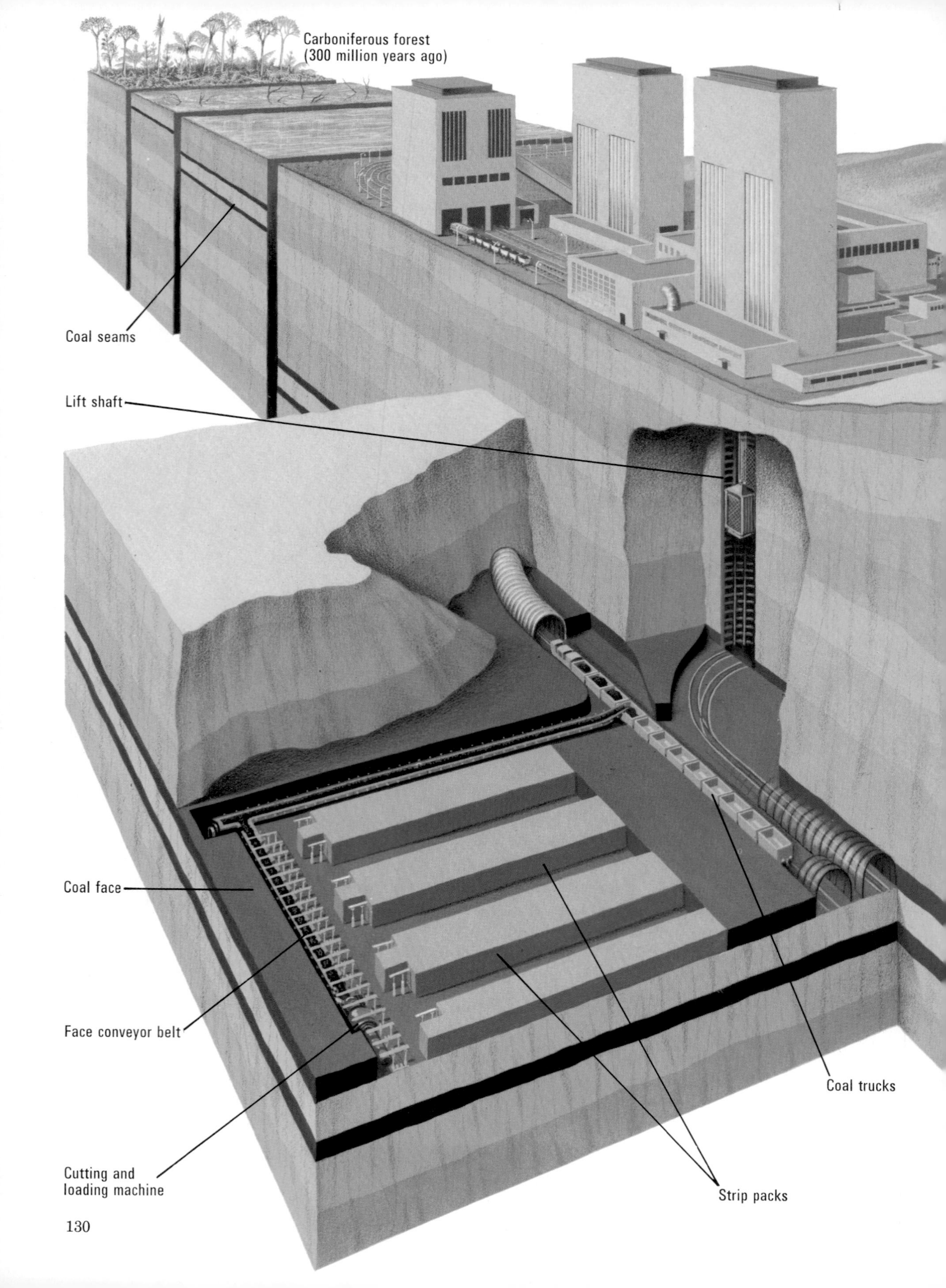
Carboniferous forest
(300 million years ago)
Coal seams
Lift shaft
Coal face
Face conveyor belt
Cutting and
loading machine
Coal trucks
Strip packs

Coal, Oil, and Gas: Formation and Exploitation

Block diagrams illustrate how man taps "fossil fuels" that originated in organisms that had lived on the earth's surface, died, decayed, and become buried and transformed within the earth's crust. Left: coal formation, and installations for mining coal from underground seams and bringing the coal aloft. Below: installations designed to drill holes into the seabed to pipe out gas and oil deposits trapped by rock structures.

Europe, Australasia, Polynesia, and the north polar lands. Wherever he settled, he learned to use the materials at hand: stone, wood, grasses, bone, metals, and precious stones. He supplemented the energy obtained from his food with fire, wind, water, and animal power. The overall effect was to create new ecological relationships, in which millions of wild animals and plants were wiped out, for as the human population grew there was less space, nutrients, and energy left for other organisms. In the long term, some of the man-made ecosystems resulted in deserts and barren, stony islands. Others were more successful, because man, ever-inventive, had learned not only how to store water for future use but also how to obtain it from underground aquifers to irrigate the land.

Man's inventiveness appears unlimited, his manifestations are everywhere. As men gathered in larger communities, they constructed artificial cliffs lined with caves in which to live and work. At night the earth is aglow with the lights of cities, releasing the energy from coal and oil.

It was but a fraction of a second ago in geological time that man's cities were built, but already they cover vast areas of land in a man-made agglomerate of stone and concrete that prevents rain from entering the porous rocks beneath. The portion of rainfall that does not evaporate runs down drains and storm sewers to the nearest river or body of water without first performing its role as the eroder and carrier of nutrients to the sea. Although the city conserves the rocks and sediments beneath its foundations, it breaks the hydrological cycle.

Urban growth is limited only by the resources that feed it. Cities absorb increasing amounts of the earth's resources without giving much that is immediately usable in return. The city, a comparatively new phenomenon—at least in terms of man's history—is spreading rapidly over the earth's crust in the temperate, subtropical, and tropical regions. Now it has even reached the northern polar wastes. With every passing year, millions of acres of productive cropland, pastures, and forests are covered by urban development.

Cities have an almost organic existence in extending and repairing themselves. They draw on the reserves of stone, sand, gravel, and clay. They take water from a vast area of land and

A vast open-cut iron-ore mine at Iron Knob in South Australia, with a miners' town beyond. Strip mining shifts huge quantities of crustal rocks; for instance, Utah's Bingham Canyon Copper Mine handles 300,000 tons of ore and overburden daily. But strip mining also disfigures the landscape by removing plants and topsoil and leaving barren, rocky terraces and dreary spoil heaps.

Waters pouring from the Aswan High Dam generate vast quantities of electricity for Egypt, and 300-mile-long Lake Nasser formed by penning back the Nile River serves to irrigate huge areas of farmland. Yet, even so, such dams may eventually create more problems than they solve, for by halting the flow of sediments they deprive regions downstream of a valuable source of nutrients.

consume huge amounts of fossil fuels, coal and oil, at an ever-increasing rate. To cope with his energy requirements, man now directs his ingenuity to finding other forms of energy. Having harnessed the energy that binds atoms together, he is now trying to emulate the sun by working out a way of controlling the hydrogen-bomb reaction to produce useful energy in the same way that the earth's star provides it. The release of all this energy in the cities is changing the composition of the atmosphere and the pattern of the air flow over and around large built-up areas; and as production and energy use increase, so does its effect. The burning of fossil fuels in dwellings, power plants, and motor vehicles pours tons of carbon dioxide into the air envelope. Tall buildings interrupt the flow of air, causing it to funnel violently between them.

In the short time he has been on the planet, man has been at least partly successful in creating an environment in which his vast numbers are more secure, but not without cost. To supply the population with food, water, and other essential materials, man has to extend his burrowing into the earth and to devise increasingly elaborate schemes for collecting and channeling water to his agricultural land and cities. Since his Neolithic ancestors first dug for flints to make their tools, man has tapped the resource wealth of the earth's crust all over the planet. He digs deep for coal, both for fuel and as raw material for the plastics and chemical industries. He also drills neat holes to release methane and petroleum. These products of organic death and decay, hitherto locked in the crustal sediments, are not only burned for energy, but also changed chemically into a host of products, some essential, some luxuries. He sends down shafts or digs great pits for copper, iron, and diamonds, and pumps hot water and compressed air down into the rocks to melt and bring to the surface the element sulfur.

As well as scarring the earth's surface, mining can also cause movements in the rock strata below. In some areas the land surface has subsided along man-made faults in the rock. In others it has dropped below the water table, forming artificial lakes. Underground water seeps into places that have never known its presence before, and the gasses trapped there are released.

The profusion of metals, precious stones, fuels, and building materials provided by the earth's crust has enabled man to bypass many of the environmental constraints that held other organisms in check. The only brake on his success is food production. He is still tied physiologically to the green plants, and even his ingenuity in producing new disease-resistant varieties of higher yield and better agricultural methods has failed to provide all the world's people with an adequate diet. Much will depend on whether industrial conversion of simple compounds into food can meet the demands for sustenance.

A shift to chemical food production still requires materials of the earth's crust and energy, and will not release man from his dependency upon the crustal rocks. Production of food in factories would mean changes in atmospheric heat distribution, interruptions in water circulation, and the redistribution of resources from their present scattered regions of origin to areas of high concentration. All will involve major shifts in the environmental balance sheet.

Irrigation schemes have already changed many parts of the world. Dams thrown across rivers prevent the nutrient sediments from reaching the valleys, deltas, and seas, to the detriment of once fertile lands and waters. The Aswan High Dam in Egypt has ended the age-old seasonal flooding and deposition of silt that made the valley of the Nile one of the cradles of civilization. And by preventing nutrients from reaching the Mediterranean, the great dam at Aswan has so reduced its productivity that the fishing that once provided food has virtually ceased.

Water storage behind dams can also alter seasonal and daily weather patterns, particularly in the warmer parts of the world. An artificial lake of several thousand acres increases the humidity of the air and the soil and gives rise to new ecological situations. Some, such as an increase of vegetation cover, counter some of the erosive activities of man, but a man-made lake also increases the territory of water-borne disease organisms. The weight of a large body of

Above: chalk cliffs near England's Beachy Head. Left: automobiles in a Los Angeles junkyard. Both contain elements concentrated by living organisms. Both may be broken down and rebuilt —the rock over many millennia, but a car in literally days, for man has the ability swiftly to recycle certain elements. He now needs to do so, for he is exhausting some ore supplies.

Exploitation of the earth's crust has allowed man's numbers to rise astronomically—at a cost counted in destruction of natural landscape and of life that shares our planet. Above: china clay spoil heaps form an unnatural skyline near St. Austell in the English county of Cornwall. Below: industrial wastes stain and may poison this water at Newcastle in southeastern Australia.

water lying over an area of faulty strata or mantle activity can cause shifts and movements in the rocks that may send out earthquake shocks. Badly planned irrigation schemes can cause the water table to rise, returning dissolved minerals to the surface to form hard sterile crusts of salt when the water evaporates.

The interruption in the flow of energy and nutrients and its redirection to human use has been achieved with remarkable speed. Such a changeover from one system to another inevitably brings problems. The overkilling of wild animals by primitive hunters, the elimination of competitors by pastoralists and farmers in the past, and the extinction and near-extinction of plants and animals at the present time is evidence that a man-centered ecosystem is full of problems. This change, when considered in evolutionary perspective, is not essentially different from the elimination of the pre-cellular molecules by the cellular organisms or the channeling of a large proportion of the energy and nutrients into the dinosaurs at the expense of other organisms, including the primitive mammals. The real differences are quantitive—in rate and the means by which man has achieved his take-over.

In principle, there is little difference between the white cliffs of Dover and automobile junk yards. Both represent the collection, utilization, and concentration of an element or group of elements by an organism, but when we consider rate and method, the human impact becomes apparent. In the case of the Dover cliffs it took countless millions of foraminifera millions of years to absorb, use, and concentrate the vast calcium carbonate deposits that form them. Then millions of years more were required for the tectonic movements within the earth's crust to raise the chalk cliffs above sea level and expose them. But in the case of the automobile dump it takes a few thousand men only a few years to collect the iron, fashion it into automobiles, use them, smash them, and scrap them. There is another difference between the two cycles, for man may reuse the iron without having to wait millions of years for it to show up again. Although the original metal resource will eventually be exhausted, man should be able to carry on his Iron Age for a long time to come, if he recycles the iron within the system.

The dangerous time for man is during this transitional phase while he is imposing his system on the existing crustal and ecological systems. Unfortunately he is damaging the ecological systems on which he still depends, leaving him no margin for error. Mining, for example, can destroy ecosystems by permitting the rapid escape of poisonous substances originally locked up in the rock strata and released slowly by environmental processes before man intervened. Now thousands of signs of his earth burrowing lie abandoned, to litter the landscape with debris and leave lakes lifeless because of poisoned waters. Industrial processes belch out wastes harmful to plants and to animals—not only to wild things but to domestic livestock and man himself. He has unleashed radioactive materials in quantities that in the course of the next few thousand years are likely to change himself, other animals, and plants. Many people have died, or will die because of these activities.

This single species has changed more of the surface of the earth's crust in 100 years than all the other life forms, both living and extinct, have done in over 600 million years. No organism, other than the first chlorophyll-bearing cells, has changed the world so completely. What the outcome will be time alone will tell, but it seems likely that man will change the environment completely to meet his own needs. The process of changing it, however, will push him farther along his evolutionary channel. Man's intelligence and flexibility should provide him with the means of surviving the environmental fluctuations that are bound to occur outside his control. His position is not quite unassailable, however, and he has to proceed with increasing caution.

The story of the earth is only half-way through. The stretch of time to come before it is finally destroyed by the failing sun is at least as long as that which has passed since the planet formed out of cosmic dust. Today's most significant development will be tomorrow's history. Through the wondrous workings of evolution, life was an inevitable consequence of the earth's chemistry, and man a consequence of that life. If man is a product of the evolutionary development of the earth, then it is not surprising that he remains dependent on it and intimately bound to it. He has already voyaged from his native planet into space; he has stepped onto another world. He has sent his instruments to the far corners of the solar system, but above all he has looked down on his own beautiful blue planet, which gave him life, continues to sustain him, and with care will continue to do so through the coming millennia.

Index

Page numbers in *italics* refer to illustrations or captions to illustrations